Reintroducing Materials for Sustainable Design

Reintroducing Materials for Sustainable Design provides instrumental theory and practical guidance to bring materials back into a central role in the design process and education.

To create designs that are sustainable and respond to current environmental, economic and cultural concerns, practitioners and educators require a clear framework for materials use in design and product manufacturing. While much has been written about sustainable design over the last two decades, outlining systems of sustainability and product criteria, to design for material circularity requires a detailed understanding of the physical matter that constitutes products. Designers must not just know of materials but know how to manipulate them and work with them creatively. This book responds to the gap by offering a way to acquire the material knowledge necessary to design physical objects for sustainability. It reinforces the key role and responsibility of designers and encourages designers to take back control over the ideation and manufacturing process. Finally, it discusses the educational practice involved and the potential implications for design education following implementation, addressing didactics, facilities and expertise.

This guide is a must-read for designers, educators and researchers engaged in sustainable product design and materials.

Mette Bak-Andersen is a Danish designer and researcher. She was educated as a designer in Barcelona and holds a PhD from the Royal Danish Academy of Fine Arts, School of Design. Since 2008 she has been working with sustainable design and explored ways to bring back materials into the design process, both in her design practice and in educational projects with design students, as well as in her doctoral research. In 2013 she founded the Material Design Lab at Copenhagen School of Design and Technology, which she directed until 2018.

"This book is a wake-up call, an appeal to educators to bring closeness to materials back into a central role in the design process and education. It is timely: the current concern for the well-being of present and future generations requires that materials be chosen in ways that are better informed about the environmental consequences of their use than at present. And at a human level, the materials of the products that surround us, if well chosen, bring an aesthetic satisfaction that is life-enhancing."

– Mike Ashby, Emeritus Professor of Materials, University of Cambridge, UK

"This book is a must-read for anyone interested in how materiality can be brought back into the center of design education."

– Mark Miodownik, Professor of Materials & Society and Director of Institute of Making, University College London, UK

Reintroducing Materials for Sustainable Design

Design Process and Educational Practice

METTE BAK-ANDERSEN

Routledge
Taylor & Francis Group
LONDON AND NEW YORK

First published 2021
by Routledge
2 Park Square, Milton Park, Abingdon, Oxon OX14 4RN

and by Routledge
52 Vanderbilt Avenue, New York, NY 10017

Routledge is an imprint of the Taylor & Francis Group, an informa business

British Library Cataloguing-in-Publication Data
A catalogue record for this book is available from the British Library

Library of Congress Cataloging-in-Publication Data
A catalog record for this book has been requested

ISBN: 978-0-367-62521-4 (hbk)
ISBN: 978-0-367-62519-1 (pbk)
ISBN: 978-1-003-10952-5 (ebk)

Typeset in Univers LT Std
by Apex CoVantage, LLC

Contents

Figures

Introduction

The motivation for writing this book was initiated 12 years ago by a growing frustration with the way we design things and how it was detrimental to the environment. As a young designer I had at the time already designed a wide variety of things from yacht interiors to an offshore transportation capsule and furniture. Most of these design processes had required collaboration and co-creation with a number of different professionals and many decisions involved with materialising the design and the production process were out of my hands. However, on several occasions they could have been in my hands or I could have influenced them, but I did not have the knowledge to do so. Consequently, none of the products that I designed during the early stages of my career as a designer can be described as sustainable. As a design educator at the time, I also felt a responsibility towards educating a new generation of designers and, thus, I started gaining knowledge on sustainability; I read the book *Cradle to cradle* (McDonough & Braungart, 2002), the Brundtland Report (Brundtland et al., 1987) together with any articles that I could find on corporate social responsibility and environmental issues. Equipped with this theoretical knowledge I started to teach about sustainability. Nevertheless, before long, it became clear that knowing about sustainability did not enable the students, nor indeed myself, to design for sustainability. From my limited experience at the time as a professional designer, I had a better understanding of how to make things than my students, but at the end of the day I was like them: good at conceptual design, able to make polished 3D drawings and build models in materials like foam, cardboard, balsawood and foam-board. There was a missing link between these competencies of the designer and the final product. One that was related to the physical realm of the design.

The obvious answer in a production chain of distributed knowledge would perhaps seem to find a person with the right knowledge to close the gap. However, the design would still be conceived without understanding or relating to the material reality it was designed for, and thus it would not just be in the hands of somebody else to redesign it for manufacturing, but also up to somebody else to make it sustainable. I experienced how the lack of material knowledge and ability to design with materials for production would leave designers without control of the fabrication of their own designs, without the ability to design for the criteria of sustainability and it would be a barrier in the exploration of new types of materials that might represent valuable and sustainable alternatives.

In an attempt to understand the complexity of designing for sustainability and to learn about materials, I started experimenting with recycled materials in my practice as a designer and conducted several experimental educational projects where the material was central from the beginning of the design process. One of the projects, which taught me most, was the design and manufacturing of 'The Story of a Lamp'. In 2010, after ten years of studying and working in Barcelona, I had returned to Denmark and established a collaboration with the Danish society for Nature Conservation. The aim of the collaboration was to use design to draw people's attention to the potential for recycling plastic waste. At the time there was a trend favouring things that were looking recycled or used, but I wanted to make an object that was timeless and clean, so that people would bring it into their home and keep it there. Therefore, I decided to make a lamp for a dining table, which at the same time could be a conversation piece for discussing recycling.

I designed the lamp; I made sketches, different form experiments and a 3D digital drawing of the final design, and, ultimately, I had the rather expensive vacuum forming mould made. With this I thought I was close to being ready for production. However, when I started sourcing the recycled plastic, which I had expected to be abundant, it turned out that it was at the time very difficult to come by in Denmark. Most industrial plastic waste was shipped abroad to be recycled and at the time plastic waste from Danish households was incinerated to produce heat or electricity. Consequently, to make the first lamps, I had to import household plastic waste from Germany. Before making the tool for vacuum forming, I had conferred with the plastic component manufacturer to ensure that the form was suitable for vacuum forming plastic and that it would fit the machine. Unfortunately, the recycled plastic behaved entirely differently from virgin plastic, which the manufacturer would normally use. The small impurities made the plastic stretch and distribute unevenly in the mould and, as a result, when the lamp was mounted, it did not hang straight.

After this followed a tiresome period of three years trying different sources of recycled plastics until I finally found a company, which produced the interior compartments of refrigerators. They had a waste fraction that was sufficiently clean, which meant the material would vacuum form evenly. Getting to production required numerous visits to the manufacturer of the lamp and a number of adjustments of the tool to suit the material, and the only reason the lamp ever went into production was thanks to a very patient and helpful manufacturer.

The lamp was well received and was exhibited in various places, among others in Denmark at Trapholt, Museum of Modern Art and Design as part of the exhibition 'Out to sea', in Germany, at the Red Dot Museum as part of the exhibition on Danish light design 'Dansk Lys: Lighting and lamp design from Denmark', and it has been used as an example of sustainable design (Plannthin, 2016). A video was made to show the costumer the process (stills from the video can be seen in Figure 0.1). Nevertheless, what the video did not show was all the trouble in trying to find a material that was compatible with the design and tool I had made, and it also did not share any of my afterthoughts.

Figure 0.1 **The Story of a Lamp**

This collage shows stills from a video documenting the process of making the lamp (Bak-Andersen, 2010). The video follows the production of the first lamps, which were made from polypropylene recycled from household waste in Germany. Later the lamps were made from polystyrene recycled from a specific Danish factory, which produces interior compartments of refrigerators. This was a cleaner material with fewer impurities and therefore it improved the quality of the lamps, but unfortunately as a result the production of the lamps depended on the waste stream from one specific source and, thus, it was not always consistent.

Retrospectively, I believe that the knowledge I gained from the difficulties in the process of making the lamp are more valuable than the lamp itself. Some of the barriers I encountered were related to the system of sustainability in Denmark; in the process of designing a recycled and recyclable product, substantial flaws within the recycling system were revealed. But most of my thoughts were related to the process of designing and making. Anyone who has been involved with the manufacturing of a design knows that it is rarely an easy process. However, had I fully understood the material I was using before I made my design and the tool for production, I could have made the process considerably simpler. If, before making the design and deciding on a manufacturing technique, I had researched the sources of recycled plastic in Denmark, I would have found different specific types of plastics. Had I explored and experimented with these, I would have understood the different materials' qualities and flaws, learned how they behaved, how they could be manipulated and which manufacturing techniques would be suitable for production.

Instead of doing this, I had based my design on a concept and an idea of material circularity, but not on the actual concrete material reality and, thus, ended up having to find a material that was compatible with the design I had made. Apart from it being very impractical, I came to feel that there was something inherently wrong with this way of designing – why would the material, which was to physically substantiate and materialise the design, be left out of the equation at any point in the design process? It was necessary to find a way to bring back the materials into the design process. Subsequently, I founded the Material Design Lab at Copenhagen School of Design and Technology, a space that was designed to experiment and design with materials, and I embarked on four years of doctoral research trying to find a way to include materials into the design process. This book contains the collected findings from my research and from my practice as a designer and design educator.

References

Bak-Andersen, M. (2010). The story of a lamp. Retrieved 06/05/2020, from https://www.youtube.com/watch?v=JECrjTkuwC0.

Brundtland, G., Khalid, M., Agnelli, S., Al-Athel, S., Chidzero, B., Fadika, L., et al. (1987). Our common future (The 'Brundtland report'). Oxford: Oxford University Press.

McDonough, W., & Braungart, M. (2002). *Cradle to cradle: remaking the way we make things*. New York, NY: North Point Press.

Plannthin, D. (2016). Fremtidens design er etisk. In L. Dybdahl (Ed.), *Dansk Design Nu*. Copenhagen, Denmark: Strandberg Publishing.

Leather made from cow stomach. When is a material culturally acceptable?

Courtesy: Copenhagen School of Design and Technology.

1 Sustainability and making

These days it is difficult to avoid being confronted with the negative environmental effects of our civilisation. Every day we are presented with gloomy reports in the media on all aspects ranging from global warming, micro-plastics and pollution to non-renewable resources running out. The changes that human life is causing to the planet are by now so comprehensive that scientists have named the geologic time in which we live the Anthropocene. It is debatable when this era began because it most likely started at different times in different places. Still, the global environmental impact of the Industrial Revolution in the early 19th century, which was caused by a combination of industrialisation and the acceleration of population growth, is generally considered a 'clear step change in the human signal' (Zalasiewicz, Crutzen & Steffen, 2012). Despite the fact that it can be difficult to see the direct effect of one(s) own life, we are all partners in crime when it comes to watching the effects of human civilisation and in particular the activity of making. As designers of the physical material world we are a central element in this story and consequently we need to decide what part we are going to play, while being fully aware of the consequences our decisions and actions might have.

An examination of current research on sustainable design reveals that there are a wealth of methods and guidelines. They vary in form and content, but usually they outline the criteria required for the final design to be sustainable (Ahmad, Wong, Tseng & Wong, 2018; De los Rios & Charnley, 2017; Dyllick & Rost, 2017; Vezzoli & Manzini, 2008). It is important to stress that the objective of this book project is *neither* to redefine sustainability *nor* to provide a singular 'true' definition or method, because sustainable development is always related to context. The objective *is* to address central aspects in the predominant design process that act as barriers to sustainability and subsequently to describe a way of designing that can enable the designer to acquire the type of material knowledge needed for practising sustainable design.

In this chapter I will address the definition of sustainability and the role of the designer within this definition and describe the knowledge required to work with sustainability.

1.1 Defining sustainability

'Sustainability' is a term that is used incessantly in multiple senses and in many different contexts, which makes it difficult to use without further explanation. Perhaps the diversity in meaning has to do with the word itself, because its etymology does not disclose what it is we have to 'sustain'. Is it human life, the ecosystem, the economy or all three? And furthermore, how should this be done? There appears to be little consensus on these matters and consequently there is a wide disparity in the term's definition. Already in 2007 it was estimated that there were more than 300 variations (Santillo, 2007). This ambiguity makes it necessary to dedicate some time to outlining how sustainability and sustainable design are understood within the context of this book in order to clarify the conceptual foundation.

Perhaps the most well-known definition is still the one offered by the Brundtland report, which states that sustainability is 'a development that meets the needs of the present without compromising the ability of future generations to meet their own needs' (Brundtland et al., 1987). The report was mainly focused on environmental aspects, but it is seen as the initial step towards defining sustainable development as a so-called triple bottom line, which apart from ecology also includes social aspects and the economy (Elkington, 1998). These aspects are more popularly known as People, Planet and Profit or the 3Ps. This framework provides a holistic view of sustainability and it has become widely accepted that sustainability comprises these three dimensions (Ahmad et al., 2018; De los Rios & Charnley, 2017; Dyllick & Rost, 2017; Sherwin, 2004; Stock & Seliger, 2016). In its foundation this approach is balanced because it recognises humans as part of ecological systems.

When considering the weaknesses of this framework, it is perhaps worth noting that the etymology of 'economics' and 'ecology' belong to the same field and share the same root: both fields deal, roughly speaking, with available resources, production, transformation, exchange, consumption and value (Findeli, 2008). Perhaps for this reason, it is not always entirely clear to what degree the beneficiary is indeed the environment. The 3Ps alone do not define to what extent or how the ecological and social aspects should be considered. These products or businesses might in some cases be 'better' categorised as 'well-intentioned eco-responses' (Reay, McCool & Withell, 2011), but frequently they should not be defined as truly sustainable (Dyllick & Muff, 2016; Reay et al., 2011; Santillo, 2007).

1.2 The shifting baseline syndrome

It would appear that humans have a tendency to accept the present situation as status quo. The people inhabiting Earth at the moment have no direct experience of what life was like before their time and, thus, cannot truly grasp the changes that have occurred, for better or for worse. A study of the way children in Houston experienced pollution concluded that 'with each generation, the amount of environmental degradation increases, but each generation takes that amount as the norm' (Kahn Jr & Friedman, 1995). Furthermore, it would seem that we also have a tendency to forget or at least grow accustomed to alterations very rapidly, e.g. accept that the fish we buy are getting smaller and smaller or that the water in the lake is no longer potable. This is called the 'shifting baseline syndrome' (Pauly, 1995). Another example of this is the population levels of insects. Research shows that in some regions there are up to 80% fewer insects than 20 years ago (Hallmann et al., 2017) and though a few adults might have noticed a difference, for children born today the present number of insects has been normalised. Unfortunately, the same acceptance of what is normal appears to be related to lifestyle. In economically wealthy societies like Denmark we have to talk to the oldest citizens to find people who have experienced a life that was not based on abundance and a growth in material wealth.

In the light of the present environmental situation, there can be little doubt that the current status is untenable. Over a longer time perspective, it might only represent a very short period, but for the people living now, it is a lifestyle we take for granted. It is the norm. This represents a big challenge to sustainability. Because even though consumers might buy more sustainable products with a slightly lighter environmental impact, as long as success for a society is based upon growth in material goods and the world's population continues to increase, the total amount of natural resources consumed will keep growing together with an increase in negative environmental impacts (Vezzoli & Manzini, 2008).

1.3 Circular economy: a system for environmental sustainability

The concrete challenges of sustainable development are at least as heterogeneous and complex as the diversity of human societies and natural ecosystems around the world (Robert, Parris & Leiserowitz, 2005). Consequently, sustainable development forces designers and other stakeholders to explore and work within complex systems and demands (Van Roon & Knight, 2003). This means that no single element of the problem can be solved in isolation: social improvements affect the environment and vice versa. An example of this is the high birth rates that follow poverty. Eliminating poverty would be a very important step against overpopulation and the environmental pressure that follows with it. Likewise ensuring that women receive an education would be another way to control birth rates, because women with higher education have fewer children (Schultz, 1994). Equally, the state of the environment also affects social aspects. Environmental degradation such as severe decline in biodiversity, climate change and imbalances in soil composition typically affect poor people more than others, because their livelihood depends on these aspects. Thus, the quality of the environment would benefit from poverty reduction (Heger, Zens & Bangalor, 2018). This is also why viewing human activities as separate from nature will ultimately be unsuccessful. Nevertheless, it is critical that decision-making is made within an ecological context and to do this we require a more defined system of sustainability.

In 2015 the 17 Sustainable Development Goals (SDG) were adopted by all the United Nations' state members (https://sdgs.un.org/goals). The separate goals are built on a holistic understanding of sustainability. However, like the 3Ps, they may permit stakeholders to take limited action within one of the goals and yet still classify the resulting products or businesses as sustainable. Naturally, it is preferable to make any change for the better, but it is insufficient when considering that we are in a situation where the side effects of our activities and the manufacturing of the human-made world have already caused a destabilisation in the Earth's ecosystem within several interconnected parameters, such as biodiversity loss, climate change and severe distortion of the nitrogen and phosphorus cycles (Rockström et al., 2009; Steffen et al., 2015). Therefore, regardless of whether the product we have designed will improve access to clean drinking water (SDG 6), will promote sustainable economic growth and employment (SDG 8) or promote gender equality (SDG 5), we must in all cases relate to the environmental impact of the design.

Circular economy is a regenerative system that provides an alternative to the extractive industrial model of *take, make and dispose* that has been predominant for centuries. In a circular economy resource input and waste, emissions and energy leakage are minimised by slowing, closing and narrowing material and energy loops. It is based upon the ideas of 'Industrial Ecology' (Ayres & Kneese, 1969), 'Performance Economy' (Stahel, 2010; Stahel, 1997), the principles of material flow and recycling of resources divided into a technical and a biological sphere presented in the book *Cradle to Cradle* (McDonough & Braungart, 2002) and the concept of regeneration of natural systems outlined in the book *Biomimicry* (Baumeister, Tocke, Dwyer, Ritter & Benyus, 2014; Kennedy, Fecheyr-Lippens, Hsiung, Niewiarowski & Kolodziej, 2015). Circular economy is regenerative by design – both at the product level and on the systemic level. Hence, design must intervene on all levels from new business models to ways of designing out waste and pollution from a product's lifecycle. It has a global perspective, addressing problems on a planetary scale that calls for a change of practice and the forms of collaboration between numerous stakeholders (Ellen MacArthur Foundation, 2015) (see Figure 1.1).

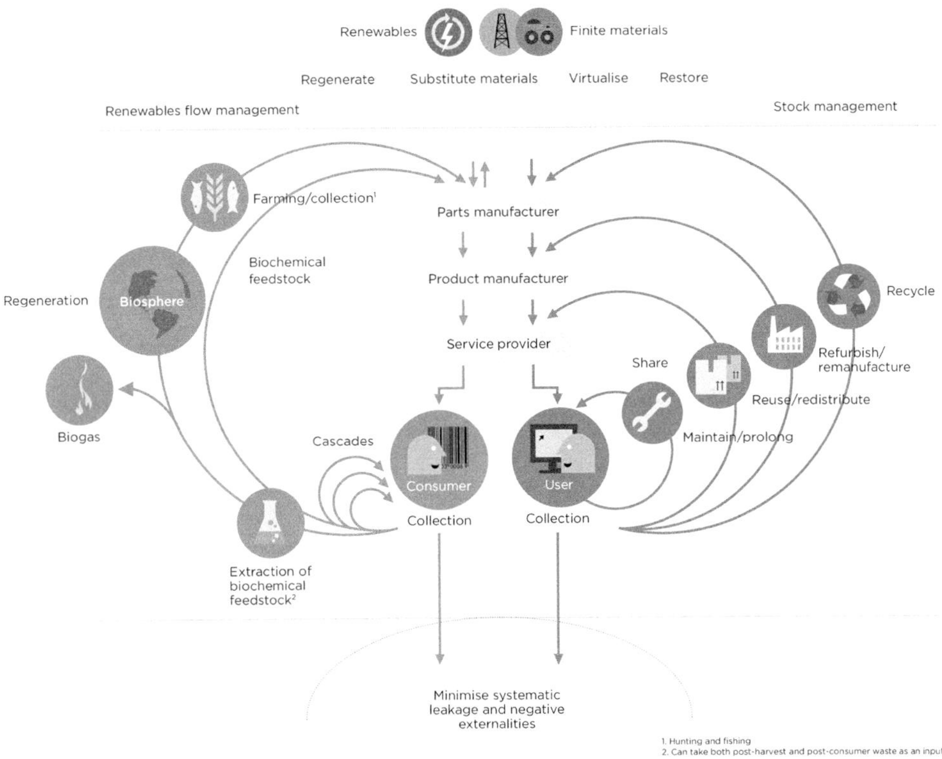

Figure 1.1 **Circular economy.**

Based on the principles from *Cradle to cradle* (McDonough & Braungart, 2002) this diagram shows a more developed system of circular economy, illustrating the continuous flow of technical and biological materials through the 'value circle'. The first step being to preserve and enhance natural capital by controlling finite stocks and balancing renewable resource flows. The second is to optimise resource yields by circulating products, components and materials in use at the highest utility, at all times, in both technical and biological cycles. The third is to foster system effectiveness by revealing and designing out negative externalities.

Courtesy: Ellen MacArthur Foundation (Ellen MacArthur Foundation, 2015).

To this day circular economy is one of the most complete systems for environmentally sustainable development that has been proposed, and perhaps therefore also the one that many governments and organisations are adopting (European Union, 2018; Giutini & Gaudette, 2003; Su, Heshmati, Geng & Yu, 2013). Still, circular economy in its present state should not be considered *the* final strategy for sustainability. The transition towards sustainability has to be an adaption-through-learning process and not least on a political level is it important to adapt and transform according to the feedback given by the environment (Vezzoli & Manzini, 2008).

Despite a general approval of circular economy in Europe, it is still far from being an integrated part of our daily practice. It is not necessary to look to research to document this, simply by taking a look around our shops, offices or most people's homes, it is plain that very few objects are being designed for circularity. Even within science the principles of material circularity are relatively new (Reay et al., 2011). This is also reflected in the rather small amount of published research on circular economy, in comparison to research about sustainability in general for example (Geissdoerfer, Savaget, Bocken & Hultink, 2017). Furthermore, some researchers point out that the main beneficiaries in a circular economy appear to be the economic actors that implement the system and not the environment, society or economy at large (Geissdoerfer et al., 2017).

Success is often measured primarily by economic growth and if this growth is tied to an increase in the fabrication of physical products, it naturally outbalances the benefits of a reduced impact of each product unit. This is also why economists who propose different alternative systems and ways to measure success are significant for defining the future of sustainability. Kate Raworth contributes a good example of this in her books on 'the Doughnut economy', in which she suggests an economic model for a development that seeks a balance where no one falls short on life's essentials, such as food, housing and healthcare, while at the same time ensuring that collectively we do not put too much pressure on Earth's life-supporting systems, on which we fundamentally depend (Raworth, 2017).

In a sustainability context, evidently it would be optimal for the physical product to vanish (Findeli, 2008) and some research concludes that designers should now be considered more as 'solution providers' and not only 'object creators' (Roux, 2011). Still, even with the greatest of attempts to avoid using valuable resources and solving problems with immaterial solutions like service design, it is hard to imagine a world in which no making of physical objects or structures takes place. Thus, as designers, at least for the foreseeable future we still need to design material objects. Circular economy may have flaws and does not explicitly offer suggestions for the social dimension, but for now it provides a system in which physical products and materials can circulate.

1.4 The role of the designer within a system for sustainability

All factors that influence a system can together be seen as a whole, but we can never understand any single aspect fully without attending to its relationship to the whole. This correlation also reflects the complexity of working with sustainable design and, at least partly, the discrepancies found in the literature between the more normative product-focused research on the one hand, and systemic thinking on the other. Achieving a balance in which the activities of human life do not destroy the environment, or ourselves, requires more than clever designers making better products. It demands changes in all parts of the system as a whole: in legislation, new business models, modifications of consumer behaviour, etc. Designers need to understand this context and this is why a systemic approach to design is considered more sustainable than approaches that focus merely on the product itself (Ceschin & Gaziulusoy, 2016).

The importance of understanding a problem fully, and the value of a systems approach to a problem, rather than accepting a narrow problem criterion, is well described in design research (Cross, 2006; Jones, 1992; Schön, 1983). To have a manageable system, convergence has to take place among all of the actors involved, i.e. all of them have to share the same idea about the results that are to be obtained and the ways in which to do it (Vezzoli & Manzini, 2008). As might be expected, studies have found that systemic transformation can be hindered when some actors involved are not aware of the role that they are expected to undertake (Senge, Lichtenstein, Kaeufer, Bradbury & Carroll, 2007). Thus, we need to consider if we as designers are aware of and prepared for this role, and, perhaps more importantly, if we are preparing new designers in such a way that they not only understand their role within the system, but also are enabled to operate at a product level within it.

There is a considerable body of research available on sustainable systems and on tools and methods for sustainable product design (Kulatunga, Karunatilake, Weerasinghe & Ihalawatta, 2015). In a survey of these methods it was found that the majority of sustainability tools and guidelines mainly concern themselves with technical design criteria, disregarding the bigger picture related to corporate strategies for sustainability (Bakker, Wang, Huisman & Den Hollander, 2014). At the same time, recent studies have found that the social dimension of sustainability is regularly neglected by sustainable product design tools, and thus conclude that these methods can only be classified as partially sustainable (Gmelin & Seuring, 2014; Kulatunga et al., 2015). In general, there appears to be a debate regarding what is the most comprehensive method, almost like a search for *the* perfect sustainable design method. This is not unlike the search for *the* true scientific design method by the 1960's Design Methods Movement (Cross, 2006). (I will discuss the general shortcomings regarding this in the following chapter.)

Some design problems can be solved with technical rationality, but many are both 'wicked' (Rittel & Webber, 1974) and 'messy' (Schön, 1983). Consequently, it is unlikely that one method will be suitable for all. Sustainability is always about context: e.g. manufacturing in Denmark requires considerably less focus on social issues than manufacturing in Bangladesh; and designing a sustainable service for car sharing requires a completely different set of skills and knowledge than designing a sustainable car. Even though the designer of services and the designer of a physical product are both part of the same system, different design problems will require different methods, and more importantly the skills and knowledge needed for them to operate are very different.

1.5 Material knowledge in sustainable design

Irel Carolina de los Rios and Fiona J. S. Charnley describe how the design of a product directly influences the way a value chain will be managed and how a systemic transformation will require fundamental changes to the practice of design. They point out that apart from proficiencies in service design and a profound knowledge of human behaviour, it is necessary for designers to acquire a deeper knowledge of materials science, engineering techniques and operational processes (De los Rios & Charnley, 2017). Although this is perhaps more correctly described as the requirements for a service designer *as well as* a product designer, it still reflects a very large repertoire of different knowledge. However, research shows that up to 80% of sustainability impacts are decided at the product design stage (Kulatunga et al., 2015; Lewis, Gertsakis, Grant, Morelli & Sweatman, 2017). This effectively means that the designer is the creator of a recipe and will inevitably make decisions that determine the qualities of the product through its lifecycle(s).

In the article, 'When matter leads to form: material driven design for sustainability' (Bak-Andersen, 2018), I compare the designer to a chef. Most people who have engaged in cooking will understand that it is essential for a chef to have a profound technical and experiential understanding of ingredients and their behaviour in different processes in order to create a new dish. Likewise, it is impossible for a designer to create the recipe for a physical object without having a profound understanding of materials and processes. When studying the steps of one of the more detailed guidelines for sustainable design (Vezzoli & Manzini, 2008), it is clear that hardly any of them can be followed without substantial knowledge about materials and how to work with them: from the minimising of material consumption, energy consumption and toxic emissions to using renewable and biocompatible resources; from optimising and improving the lifespan of materials and products to designing for reuse, disassembly and recycling. Most of these focus on technical properties of the materials, but knowing how to make use of the experiential qualities of a material can be equally important for a product (Karana, Pedgley & Rognoli, 2013) – also within the context of sustainability. The aesthetics, the cultural perception and the perceived value of a material are all intangible subjective factors that create meaning and emotional attachment to a product, which will contribute to ensuring maintenance and durability.

Likewise, it might seem that social and economic factors are mostly connected to the system of sustainability and not as much to the materiality of the product. However, economy and ecology are focused on several of the same aspects and although the shared interest is not necessarily reflected in shared objectives, it still means that knowledge of materials and processes is just as important for economic reasons. Even when addressing social sustainability this is essential, because tracing the material used in a product, from the origins of acquiring the raw material, through production to the end of its life, will give access to considerable information about the people involved in the process and their working conditions.

Consequently, sustainability obliges designers to work on two levels: product and system. Designers must *understand* the part that they play in the dynamic whole, but they must also be able to *operate* at product level. To do this, designers must have a deep knowledge of materials and how to work with them. As I will describe in the following chapters, this is not a type of knowledge that can solely be acquired through lectures or reading and so it challenges not only how we understand the role of materials within the design process, but also how we learn to design. The aim of this book is to explain why materials must be central in design, to suggest how working with materials for production can be combined with a contemporary design process and how this can be included into educational practice.

References

Ahmad, S., Wong, K. Y., Tseng, M. L., & Wong, W. P. (2018). Sustainable product design and development: A review of tools, applications and research prospects. *Resources, Conservation and Recycling*, *132*, 49–61.

Ayres, R. U., & Kneese, A. V. (1969). Production, consumption, and externalities. *The American Economic Review*, *59*(3), 282–297.

Bak-Andersen, M. (2018). When matter leads to form: material driven design for sustainability. *Temes De Disseny: Nueva Etapa*, *34*, 10–33.

Bakker, C., Wang, F., Huisman, J., & Den Hollander, M. (2014). Products that go round: exploring product life extension through design. *Journal of Cleaner Production*, *69*, 10–16.

Baumeister, D., Tocke, R., Dwyer, J., Ritter, S., & Benyus, J. M. (2014). *Biomimicry resource handbook: a seed bank of best practices*. Missoula, MT: Biomimicry 3.8.

Brundtland, G., Khalid, M., Agnelli, S., Al-Athel, S., Chidzero, B., Fadika, L., et al. (1987). Our common future (Brundtland report). Oxford: Oxford University Press.

Ceschin, F., & Gaziulusoy, I. (2016). Evolution of design for sustainability: from product design to design for system innovations and transitions. *Design Studies*, *47*, 118–163.

Cross, N. (2006). *Designerly ways of knowing*. New York, NY: Springer.

De los Rios, I. C., & Charnley, F. J. S. (2017). Skills and capabilities for a sustainable and circular economy: the changing role of design. *Journal of Cleaner Production*, *160*, 109–122.

Dyllick, T., & Muff, K. (2016). Clarifying the meaning of sustainable business: introducing a typology from business-as-usual to true business sustainability. *Organization & Environment*, *29*(2), 156–174.

Dyllick, T., & Rost, Z. (2017). Towards true product sustainability. *Journal of Cleaner Production, 162*, 346–360. Retrieved from http://www.sciencedirect.com.ez-kab.statsbiblioteket.dk:2048/science/article/pii/S0959652617311423.

Elkington, J. (1998). Partnerships from cannibals with forks: the triple bottom line of 21st-century business. *Environmental Quality Management, 8*(1), 37–51.

Ellen MacArthur Foundation. (2015). Towards the circular economy: business rationale for accelerated transition. Cowes, Isle of Wight, UK: Ellen MacArthur Foundation.

European Union. (2018). *Circular economy.* Retrieved 03/01/2018, from http://ec.europa.eu/environment/circular-economy/index_en.htm.

Findeli, A. (2008). Sustainable design: a critique of the current tripolar model. *The Design Journal, 11*(3), 301–322.

Geissdoerfer, M., Savaget, P., Bocken, N. M., & Hultink, E. J. (2017). The circular economy: a new sustainability paradigm? *Journal of Cleaner Production, 143*, 757–768.

Giutini, R., & Gaudette, K. (2003). Remanufacturing: the next great opportunity for boosting US productivity. *Business Horizons, 46*(6), 41–48.

Gmelin, H., & Seuring, S. (2014). Achieving sustainable new product development by integrating product life-cycle management capabilities. *International Journal of Production Economics, 154*, 166–177.

Hallmann, C. A., Sorg, M., Jongejans, E., Siepel, H., Hofland, N., Schwan, H., et al. (2017). More than 75 percent decline over 27 years in total flying insect biomass in protected areas. *PloS One, 12*(10), e0185809.

Heger, M., Zens, G., & Bangalor, M. (2018). *Does the environment matter for poverty reduction? The role of soil fertility and vegetation vigor in poverty reduction.* Washington, D.C.: The World Bank.

Jones, J. C. (1992). *Design methods.* Hoboken, NJ: John Wiley & Sons.

Kahn Jr, P. H., & Friedman, B. (1995). Environmental views and values of children in an inner-city black community. *Child Development, 66*(5), 1403–1417.

Karana, E., Pedgley, O., & Rognoli, V. (2013). *Materials experience: fundamentals of materials and design.* Butterworth-Heinemann.

Kennedy, E., Fecheyr-Lippens, D., Hsiung, B., Niewiarowski, P. H., & Kolodziej, M. (2015). Biomimicry: a path to sustainable innovation. *Design Issues, 31*(3), 66–73.

Kulatunga, A., Karunatilake, N., Weerasinghe, N., & Ihalawatta, R. (2015). Sustainable manufacturing based decision support model for product design and development process. *Procedia CIRP, 26*, 87–92.

Lewis, H., Gertsakis, J., Grant, T., Morelli, N., & Sweatman, A. (2017). *Design environment: a global guide to designing greener goods.* Abingdon-on-Thames, UK: Routledge.

McDonough, W., & Braungart, M. (2002). *Cradle to cradle: remaking the way we make things.* New York, NY: North Point Press.

Pauly, D. (1995). Anecdotes and the shifting baseline syndrome of fisheries. *Trends in Ecology & Evolution, 10*(10), 430.

Raworth, K. (2017). *Doughnut economics: seven ways to think like a 21st-century economist.* Chelsea, VT: Chelsea Green Publishing.

Reay, S., McCool, J., & Withell, A. (2011). Exploring the feasibility of cradle-to-cradle (product) design: perspectives from New Zealand scientists. *International Journal of Sustainable Development*, *4*(1), 36.

Rittel, H. W., & Webber, M. M. (1974). Wicked problems. *Man-made Futures*, *26*(1), 272–280.

Robert, K. W., Parris, T. M., & Leiserowitz, A. A. (2005). What is sustainable development? Goals, indicators, values, and practice. *Environment: Science and Policy for Sustainable Development*, *47*(3), 8–21.

Rockström, J., Steffen, W., Noone, K., Persson, Å., Chapin III, F. S., Lambin, E. F., et al. (2009). A safe operating space for humanity. *Nature*, *461*, 472–475.

Roux, C. (2011). The new meaning of product design? *Design Management Review*, *22*(4), 22–25.

Santillo, D. (2007). Reclaiming the definition of sustainability. *Environmental Science and Pollution Research-International*, *14*(1), 60–66.

Schön, D. A. (1983). *The reflective practitioner: how professionals think in action*. Abingdon-on-Thames, UK: Routledge.

Schultz, T. P. (1994). Human capital, family planning, and their effects on population growth. *The American Economic Review*, *84*(2), 255–260.

Senge, P. M., Lichtenstein, B. B., Kaeufer, K., Bradbury, H., & Carroll, J. S. (2007). Collaborating for systemic change. *MIT Sloan Management Review*, *48*(2), 44.

Sherwin, C. (2004). Design and sustainability. *The Journal of Sustainable Product Design*, *4*(1–4), 21–31.

Stahel, W. (2010). *The performance economy*. New York, NY: Springer.

Stahel, W. R. (1997). The service economy: 'wealth without resource consumption'? *Philosophical Transactions of the Royal Society of London A: Mathematical, Physical and Engineering Sciences*, *355*(1728), 1309–1319.

Steffen, W., Richardson, K., Rockstrom, J., Cornell, S. E., Fetzer, I., Bennett, E. M., et al. (2015). Planetary boundaries: guiding human development on a changing planet. *Science*, *347*(6223), 736.

Stock, T., & Seliger, G. (2016). Opportunities of sustainable manufacturing in industry 4.0. *Procedia CIRP*, *40*, 536–541.

Su, B., Heshmati, A., Geng, Y., & Yu, X. (2013). A review of the circular economy in China: moving from rhetoric to implementation. *Journal of Cleaner Production*, *42*, 215–227.

Van Roon, M., & Knight, S. (2003). *Ecological context of development*. Oxford, UK: Oxford University Press.

Vezzoli, C., & Manzini, E. (2008). *Design for environmental sustainability*. New York, NY: Springer.

Zalasiewicz, J., Crutzen, P. J., & Steffen, W. (2012). The Anthropocene. In F. Gradstein & M. Schmitz (Eds.), *The geologic time scale* (pp. 1033–1040). Amsterdam, Netherlands: Elsevier.

A student studying the collection of natural materials at the Material Design Lab.

Courtesy: Copenhagen School of Design and Technology.

2 Materials in design education

Non-designers are often surprised to find that designers do not necessarily have a substantial knowledge about materials. However, examining the design process confirms that it appears to have been predominantly immaterial for some decades, which may partly explain why knowledge about materials and how to work with them is not automatically a part of a designer's toolbox. Furthermore, learning to design with materials requires more than formal knowledge and 'knowing about' the subject: it demands a kind of embodied knowledge of materials and processes that enable designers to practise creatively. To achieve this involves a certain type of educational practice and the question is how this aligns with contemporary design education.

Materials have been central in the design process before, so in order to understand ways to bring them back and how materials, making and the design process are interrelated in a contemporary context, it is valuable to explore the history of materials within design and design education, and more specifically to study how didactic ideologies and definitions of design may have affected the constituent parts of how we see the design profession today. Research has provided the design profession with a considerably more solid theoretical foundation than it had 50 years ago. Nevertheless, as has already been suggested, there are elements within an immaterial design process that can be counterproductive when designing physical sustainable objects.

The aim of this chapter is not to provide an exhaustive account of design education history, but rather to follow the role of materials and making within educational curricula across different periods of time. This information can help provide an understanding of why materials and making in some historical epochs have been considered central to the creative process of designing physical objects and why in other periods they have been almost ignored. Apart from a more general historical overview (mainly based upon data from Northern Europe), which is illustrated in Figure 2.1, the focus will be on three examples, which are particularly interesting, namely: the Birmingham School of Art in the UK at the end of the 19th century, the Bauhaus in Germany in the 1920s and Danish modernism in the 1950s.

2.1 The dispute about materials within design education

The transition that has taken place in design, from before the Industrial Revolution when it was viewed as being an integral part of the crafts, to a profession in its own right as it is today, has been related to the transformation in design education. Traditional craft was centred on learning about a specific material or group of materials and focused on the manual skills and techniques required for their manipulation. This was done through apprenticeship, where the apprentice mimicked the work of the master and acquired the knowledge and skills necessary through copying and repetition. Designs typically evolved over generations through small alterations. Perhaps the first sign of design being separated from craft, and also the first sign of materials being disconnected from the design process, was the act of drawing. Draftsmanship had originally been an aid to

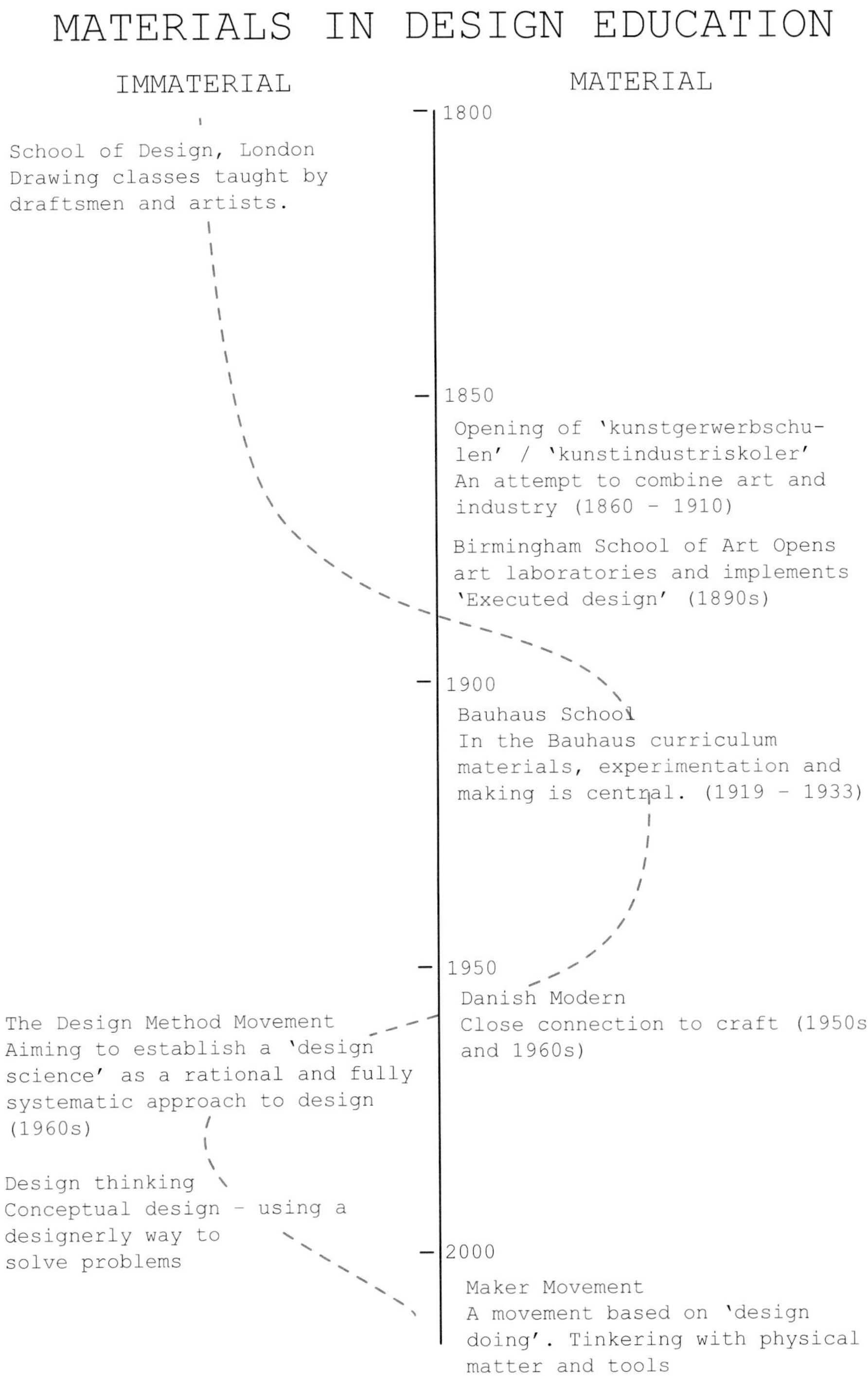

Figure 2.1 **Materials in design education.**

Based on the examples provided in the text, this illustration provides an overview of how the focus of designing has been altering over the last 200 years, between being an activity based on a material reality to being more immaterial and conceptual.

the master craftsperson; however, the status of the professions was changing and both the act of drawing and the draftsman were becoming more respected during the 19th century (Romans, 2005). This was reflected in the publication of a large number of instruction manuals for drawing, most having a technical or elementary focus, but a few had a more artistic or theoretical approach such as *The drawing book of the Government School of Design* by William Dyce (Dyce, 1843) or *The elements of drawing* by John Ruskin (Ruskin, 1857).

Nevertheless, drawing initially turned out to be rather controversial within the process of establishing design education. In the UK, this was reflected in early attempts at defining a curriculum for the new government sponsored School of Design in London during the late 1830s. The school, which much later became the Royal College of Art, was initially set up to further the knowledge of arts and the principles of design to the people – especially to the 'manufacturing population'. The aim of this was not a division of labour, but rather the opposite: to combine art with industry in order to achieve better-quality products (Cole & Redgrave, 1849). But, despite the fact that there appeared to be an agreement that craftspeople would benefit from learning to draw, it was unclear how this should be done and if it should have an artistic or a more technical focus. Such a disagreement might seem a minor detail, but at the time it was a considerable conflict. The historian Lara Kriegel describes how this dispute at the school became the staging ground for a larger discussion on aesthetic principles and artisanal practices that would later inform the project of design reform in the decades to come (Kriegel, 2007).

By 1849, there were 21 schools of design in the UK, which more than 15,000 men and women had attended. There are indications that the schools did succeed, to some extent, in training workers in art skills for manufacturing during the early years (Schmiechen, 1990). But, Henry Cole, who at the time facilitated many innovations in education, was an early critic of the design schools, because they did not include practical skills or working with materials (Cole & Redgrave, 1849). Kriegel also states that these schools initially suffered a lack of success, perhaps caused by a conflict of interests or simply down to the fee and opening hours of the schools, which were not easily compatible with the daily life of the working class (Kriegel, 2007). From the artists' point of view, the schools were perhaps not ideal either; Gottfried Semper, a prominent contemporary design-theorist, wrote that the 'influence from the heights of academic art generally lacks a practical foundation, as skilled and highly talented designers and model-makers are not metalworkers, potters, carpet weavers, and goldsmiths – as was often the case before the academies separated the arts' (Semper, 2004, p. 76).

Already in 1849, in the first issue of the *Journal of Design and Manufactures*, it had been pointed out that during the process of designing and manufacturing an object, the person making the drawings only rarely also executed them (Cole & Redgrave, 1849). This division of labour could, as Cole suggested, be put down to a deficiency and inadequacy in the curriculum of the schools, which mostly focused on drawing skills, but in hindsight it could also partly be an effect of an advance in technology and the subsequent specialisation of the workers (Adamson, 2013).

The same journal, which only ever published six volumes, included physical material samples in some issues. This is an interesting detail, because it could suggest that the bodily experience of the material was still valued as something that could not necessarily be replaced by a description in words and images (Library University of Glasgow, 2018). The quest for better quality in materials and making was equally an ambition of designer and social activist William Morris, who was perhaps the most important figure in the Arts and Crafts Movement. He wanted to restore craftspeople to their rightful place in history and make them 'once again the masters of their own hands' (Morris, 1888).

Morris was inspired by the philosopher and socialist Karl Marx, and wrote like many others about the separation between craft and trade, but his visions for the 'revival of the handicraft' included an element of nostalgic idealism. His picturesque work, which was, as it happens, not always executed by himself, primarily manifested itself as luxury goods and continues to be highly valued to this day. But to many people, the Arts and Craft Movement is seen as the essence of craft and the ideas of the movement became highly influential for many designers and design educations to come (Adamson, 2013).

Perhaps the dispute about the accomplishments of the design schools of the 19th century reflects diverse expectations as to what these should actually achieve. Thomas Wright, one of the few craftsmen writing at the time, believed that the schools should provide a general education for the mind of the worker (Wright, 1868). However, most literature from the period suggests that the ambition was either to instil art into the growing industry in order to achieve better quality and more aesthetic products, or to follow the more artistic and almost anti-industrial ideas of Ruskin and Morris and provide practical and aesthetic design education for craftspeople (Adamson, 2013; Morris, 1888; Ruskin, 1857). Still, most 19th-century art and design educations in the UK were not initially founded on the material knowledge from craft. The focus was on drawing, and classes were taught by painters or draftsmen, not by master craftspeople. Perhaps for this reason, a lack of contact with materials for which the designs were intended was a frequent feature at this point in time and this in turn meant that the graduates from the design schools were not prepared for designing products for manufacturing (Swift, 2005).

These early disputes in building design education are interesting because they touch upon the core of the question regarding the type of knowledge and skills required to design a physical object: Is design an abstract act of the mind imagining and visualising the idea of a form, or does it require practical skills and the experience of working with 'real' materials? This debate was central to discourse during the late 19th century, but, as we will see in the following subsections, it appeared to be timely a century later and continues to be today.

2.2 From drawing to 'executed design'

Already by the middle of the 19th century, Cole had pointed out that a lack of knowledge about materials and skills in making made it difficult to design for production. Finally, in 1893 the Birmingham School of Art, which up to that point had focused on drawing, made fundamental alterations to both curriculum and spaces by opening a series of 'art laboratories' that provided the facilities and equipment for executing designs in real materials. These laboratories were frequently staffed by members of the Arts and Crafts Movement and facilitated a range of activities that involved materials and making (Swift, 2005). The Birmingham School of Art became one of the most significant art schools of the period, providing extensive, hands-on experience in all art, craft and design areas, which was coupled with an emphasis on drawing from nature and from memory. This was only reinforced by the new headmaster in 1903, Robert Catterson Smith, who not only believed that drawing from real natural objects should be compulsory, but also that design should be carried out using real materials, not just on paper. Practical designing was central and was to be included from an early stage of the student's experience. The Birmingham School of Art did not just develop new ideas concerning the role and purpose of drawing regarding memory and imagination; they established a 'new' method of learning design competencies by using the actual materials. They called the method 'executed design'. Other schools of art soon followed suit, e.g. the Central School in London and Leicester School of Art. By the second decade of the 20th century, most schools in the UK had opened practical workshops for designing (Romans, 2005).

This period of the Birmingham School of Art, which continued until 1920, was perhaps the closest to an Arts and Crafts practice education that has existed. The 'executed design' method developed at Birmingham is interesting because it made the students learn through a design process in which experimentation, development and prototypes were carried out using 'real' materials, 'real' meaning the material for which the design was intended and not a substitute material for model-building purposes. Most design schools in the beginning of the 20th century had accepted that the action of imagining and representing ideas visually in drawing was insufficient when designing physical objects for manufacturing, and that it was indeed necessary to include practical work using the real materials (Romans, 2005). As some of the examples below will show, variations of this approach to design did remain in some places for almost half a century. Still, in the transition to a fully industrial production most design curricula appeared to slowly lose contact with practical making and material experimentation.

The ambition with this book is not simply to return to the craftsmanship of old times, but to embrace technology and industrial production. Thus, it is relevant to look at the more influential design curricula of this transition period and try to understand why they eventually rejected working with real materials during the design process. In the beginning of the 20th century, when the design and manufacturing were no longer automatically carried out by the same person and industrial production was taking over, the development of '*Kunsthandwerk*' appeared, differentiating itself as the *art* work of the hand (Muthesius, S., 1998). In Germany, the architect, Hermann Muthesius argued that the dichotomy of art works and factory production was indeed a false one (Muthesius, 1901). He was among the founders of Deutscher Werkbund, an association of craftspeople that included artists, architects, designers and industrialists. It became an institution that is often referred to as a precursor to the Bauhaus School (Adamson, 2013). This focus on establishing a closer connection between art and industry seemed to prevail both in Northern Europe and in the USA. During the latter part of the 19th century, many *Kunstgerwerbschulen* – art trade schools – were opened across Germany as well as *kunstindustriskoler* – art industry schools – in Scandinavia. Many of these were closed down before the Second World War, but some stayed active for more than a century. This somewhat uneasy relationship between art and industry seems to be the field in which modern product design was born. Findeli describes that it was 'as if design was summoned to absorb the shock of industrialisation with the initial tools being aesthetics' (Findeli, 2001).

2.3 Handmade for mass production

Perhaps the most influential design school to this day is the Bauhaus School. Despite the strong connection with the Arts and Crafts Movement, the architect and initial director of the school, Walter Gropius, was not an opponent of the machine. Considering some of the more technically advanced designs that were produced by designers, architects and artists affiliated with the school, it would appear that this was indeed a place where industry and art found common ground. Yet, the initial manifesto, issued in 1919 by Gropius, was based upon the return to craft. He wanted to break with the separation and class division between fine arts and craft and in his manifesto urged architects, sculptors and painters to return to craft (Bullock & Conrads, 1975). More specifically, Gropius did this by inviting some of the most progressive teachers, artists and architects of the period to join him as faculty at the school. Despite the fact that the Bauhaus only existed for 14 years, it made a great impact on the definition of the design profession and continues to inspire design educators today.

In its short lifetime the school had three directors with very different ideas on didactics and politics and a number of teachers only stayed for a short period due to the resulting internal disagreements. Furthermore, the Bauhaus moved three times and was increasingly challenged by the political situation in Germany, which made life difficult for the school and ultimately led to its closure in 1933 (Droste, 2002). Consequently, the Bauhaus was never one singular thing, but could perhaps better be described as a constantly changing laboratory for creative productive education.

In relation to the position of materials within design education, the Bauhaus stands out. Gropius created a formative curriculum in 1922, which encompassed a gradual technological evolution from the most basic tools and techniques to the use of advanced machines – all subjects in which the student studied the reality of the matter through experience. It started with the obligatory *Vorkurs*, the preliminary course that was designed by, and in the early years also taught by, Johannes Itten, which encouraged self-expression and experimentation with materials. Subsequently, the students had to learn about colour, composition, construction, space, materials and tools, as well as study nature and matter. This would prepare them to enter a specialisation, which at the time could be clay, stone, wood, metal, textiles, colour or glass, where they would be exposed to more technical and theoretical elements. This finally led them to the main objective of the programme: *Bau* (meaning 'build' or 'construct') (Droste, 2002). It was a progression of learning and a way of designing that was based upon the education reformer John Dewey's ideas of learning by doing (Ascher, 2015), and the methodology of Itten's *Vorkurs* in particular was inspired by María Montessori's work with children in which the transformation of matter is processed through intuition, imagination and creativity (Donoso, Miraudа & Jacob, 2018).

Though Gropius had a strong position in determining the direction of the school initially, particularly during his time as director, the school changed from being centred on arts and craft to being more focused on industry. Gropius's recognition of the industry as a significant influence on the design approach at the Bauhaus was reflected in the pamphlet for the 1923 exhibition, in which he declared that art and technology was the new unity (Droste, 2002). The transition of the school is exemplified in an interview with textile designer and artist Anni Albers in 1968. She was initially a student and later a teacher at Bauhaus. When asked if the connection to the industry was strong, she answered:

> Not in the early years. This idea of industry gradually developed and it really came on much stronger after Gropius left. Because, in the early years, there was a dabbling in a kind of romantic handicraft where you made beautiful pillowcases which – well, you couldn't wash them, perhaps you couldn't sit on them. And these tablecloths in very brilliant and bright colours. But this wasn't what was suited for industry. They couldn't make it in a hundred different threads and colours and so on. And also it wasn't satisfying because it was an over-subjectifying of something that wasn't worth it. When you have a tablecloth that is so active you can't put a plate on that tablecloth, you can't put a vase with flowers on it, it was far too dominating. And we only very gradually turned to this what I explained to you coming with this very excellent sentence from Klee to these 'serving objects'.
>
> (Albers, 1968)

Her answer reflects the school's transition of focus from fine arts to what Klee referred to as 'useful' arts and the acceptance of designing for industrial production. The connection with industry became even stronger when Hannes Meyer took over as director in 1928, which provoked some alterations to the curriculum (Droste, 2002), but the material foundation from craft was present in the workshops throughout the existence of the school.

Despite the fact that the Bauhaus was closed down in 1933, ideas, experiences and teaching methods from the school were not entirely lost. Many aspects were continued and spread in different places around the world by former Bauhaus teachers and students who left Germany because of the political situation and the outbreak of the Second World War. Some of the more famous examples were Walter Gropius and Marcel Breuer, who became teachers at the Harvard Graduate School of Design. Anni and Josef Albers went on to teach in Black Mountain College in Chicago, and in the 1950s J. Albers became the director of the design department at Yale. In 1937, Moholy-Nagy established the New Bauhaus School in Chicago (closed down the following year, but re-established as the Institute of Design in 1939) where he continued the experimental approach from the Bauhaus (Findeli, 1990, 2001). And not least Max Bill, a former Bauhaus student, co-founded the Hochschule für Gestaltung in Ulm (HfG) in 1953. Initially the HfG explicitly claimed the heritage of the Bauhaus, but, soon after, new tendencies provoked a more scientific and technical approach to design (see below).

In general, the 1950s stand out as a time in which craft and industry/technology were not necessarily seen as being in opposition, and thus the materials for which the designs were intended were still in the hands of the designers during the design process. The same balance was, to some degree, traceable between craft and theory. In this period, the American designer Don Wallance described design as being a conceptual process in its essence but emphasised that in actual practice design cannot readily be considered separated from craftsmanship. He saw design as interwoven with craftsmanship throughout the creative process, both when the object was handmade and when it was intended to be produced by a machine (Wallance, 1956). Although there was little evidence of a theoretical foundation, in a speech at the Asilomar Conference in 1957 in California, Marguerite Wildenhain, a ceramicist and former Bauhaus student, argued that there was no reason to look at the crafts and industry as enemies, because good design, whether it was a unique handmade object or a mass-produced one, depended upon the same basic qualities in the designer, namely:

1 The fusion of material and object; that means extensive knowledge of what materials are available and suited, how they can be treated so as to get the most characteristic and alive expression and quality.
2 The competent use of techniques, processes or methods, of tools or machines for the one special purpose.
3 The solving of all problems of use and function adequately and masterfully, but not only on the surface and roughly so, but in all details (knobs included!).
4 Creative use of lines, colors, volumes, tensions, form and decor. And last but not least, artistic integrity.

(Adamson, 2010, p. 571)

An example that stands out, representing this balance in which the design was still 'handmade' but intended for industrial production, is Danish design during the 1950s and 1960s. Danish Modern is often described with reference to its aesthetic style and expression; however, on closer inspection Danish modernism is perhaps more precisely defined by its ideology and the designers' profound understanding of materials (Mussari, 2016). This movement did not appear out of nowhere, but rather it can be traced back more than 50 years. From an educational point of view, perhaps the most influential changes were in 1924, at the same time as the Bauhaus School was being established, when the designer Kaare Klint became the director of the Royal Danish Academy's Furniture School. The development of Danish design education had until then, as in other places, been a fluctuation of fine arts and craft meeting and separating. And as elsewhere, technical skills, making and working with the materials for which the designs were intended were largely absent during the latter part of the 19th century, but at the beginning of the 20th century this changed.

The masters of Danish design in the 50s and early 60s, the period also known as Danish Modern, were designers like Hans Wegner, Poul Kjærholm, Poul Henningsen, Børge Mogensen, Arne Jacobsen and Verner Panton. Most of them had a strong connection to craft and several of them were even trained as craftspeople. Wegner, Kjærhjolm and Mogensen finished their apprenticeships as cabinetmakers before starting their educations as architects or designers. Verner Panton and Arne Jacobsen were trained as bricklayers before becoming architects, and – intriguingly – the most intellectual one of them, Henningsen, was 'only' ever educated in a technical school.

Their practical experience in making and handling materials gave them an understanding of processes and materials, both technically and experientially, which they brought into the development of their designs. Interestingly, some of these designers had the ambition to create democratic design, which might primarily be related to the general political tendencies in Scandinavia at the time (Hollingsworth, 2009). Nevertheless, it meant that they were designing not just on a product level, but also with some degree of understanding of its relation to a larger system. This resulted in design processes that in many cases were still based on the principles from craft, but that were combined with the ideology of democratic design, functionalism and ergonomics.

Peter Dormer proposes that people who trained in the 50s and 60s had two types of knowledge. Both the personal knowhow of craft experience and the ability to design for a more advanced production chain of distributed knowledge (Dormer, 1997). Nevertheless, the success of Danish Modern can likely be attributed to various factors, such as acquisition power due to economic growth after the Second World War, or simply the aesthetics that characterised the movement. But, as Mussari points out, perhaps the most important reasons for the popularity of these designs were (and continue to be) the ability to design for mass production without losing trace of the human factor and the quality of material expression for which craft is renowned (Mussari, 2016).

2.4 From hands to head

In the late 1990s, Peter Dormer described how design has been in a gradual process of divorcing itself from craft since the 1920s, which resulted in the separation of 'having ideas' and 'making objects' (Dormer, 1997). The strong connection between design and craft, which involved material exploration and making, slowly subsided and gave way to a more rationalistic and intellectual approach to design, dominated by a strong confidence in science and technology. Already in 1956, a conference on art education in Bretton Hall in the UK became a forum for discussions on ideological and methodological changes to the creative process. The role of the traditional crafts were critically analysed and their relevance and compatibility with the world of science and technology were questioned (Yeomans, 1988). At the conference, the artist Richard Hamilton presented a pedagogy, similar to the ideas of the artist Marcel Duchamp, in which he argued that art education should be put to the service of the mind. Both artists rejected the sensual and physical aspects of art, wishing their work to be determined by procedures and processes that were clearly the product of the mind (Romans, 2005).

This change was also noticeable in the approach to design at the HfG in Ulm. Initially, the school's curriculum was very similar to that of the Bauhaus, but after a few years Tomás Maldonado put forward ideas of a new educational philosophy that was based upon 'scientific operationalism' (Maldonado, 1960). Hamilton was interested in this development at the school because he saw it as an example of a rational method of a pedagogy that led to the exclusion of self-expression. He believed that the adult student was beyond the need for self-expression, and its 'nurturing in higher education was a misapplication of Montessori in the adult realm' (Romans, 2005). Despite its official tribute to the Bauhaus, this pedagogy and didactic approach was in many respects the antithesis of the *Vorkurs* taught by Itten at the Bauhaus.

During the 1960s there was a general movement towards establishing a design science. The overall ambition was to define an explicitly organised, rational and fully systematic approach to design. Buckminster Fuller was one of the forerunners and at the 1965 seminar on 'The design method' at Birmingham College of Advanced Technology this movement became more defined. The aim of the conference (and several other conferences during the 60s) was to devise *the* design method – a single rationalised method that could be compared to what was understood as 'the scientific method' (Cross, 2006). During the 1960s and 70s several design methods were proposed (Alexander, 1964; Archer, 1964; Broadbent, 1979; Jones, 1966; Rittel, 1984). There was an actual Design Methods Movement and a Design Methods Group was established at Berkeley University. Most of the methods from this period tried to reduce the more ill-defined craft and practice-based design process to a more rational scientific approach.

Despite the fact that this rather rigid and computational approach was later criticised, including by the authors themselves, some elements from contemporary design processes can be traced back to this movement (Rith & Dubberly, 2007). In 1969, Herbert Simon published the book *The sciences of the artificial* (Simon, 1969). The book did not specify a design method as such, but it followed the dominant design philosophy of the time, by suggesting that design should be driven by computational and technical rationality. It was also this book that stipulated the famous definition of design as being 'changing existing situations into preferred ones'. A view which opened up for an understanding of design as a much broader activity, which furthermore could be relevant to many different professions. This understanding of design as a tool to solve many different types of problems formed part of the early foundation of what has become known as 'Design Thinking'.

The book *The reflective practitioner* (Schön, 1983) also contributed to our understanding of design thinking with the idea that design was indeed an activity that could be used for problem solving in many situations. However, the philosopher Donald Schön highlighted that a strictly scientific approach would only ever be apt for solving technical and very well-defined problems. He saw the problems of real importance as being messy and ill-defined and concluded that they required a different method in order to be solved – what he called 'reflective practice'. This strategy paved the way for a more practical and reflective approach to design and problem solving, one that was not necessarily based upon technical rationality, but that allowed for more open experimentation, which also included tacit knowledge and intuition. In the 60s these attributes were considered unscientific, but were later described by Nigel Cross as inherent to a 'designerly' way of knowing (Cross, 2006).

Design Thinking can hardly be described as one thing, but in many ways it reflects the design philosophy of the following period, in which the design process predominantly became an activity of the mind. Problem solving by means of design can be incredibly complex and almost inevitably requires extensive considerations of ideas and methods from a broad array of professions outside of the field of design. In principle, processes like Design Thinking allow for this complexity; it embraces a 'designerly' approach and has furthermore been very suitable for the Bologna process (Bologna Declaration, 1999), which called for a transition of design education into a university degree, entailing a much stronger theoretical foundation of the curricula. Consequently, it is not surprising that Design Thinking as a method continues to be a popular approach in design education.

The architect and design researcher Bryan Lawson describes how design education has experienced a transition where it has progressively moved from the workplace, where one is in contact with the making of things, into the college and university studio (Lawson, 2006). Nevertheless, whereas Design Thinking and similar approaches might be suitable for problems that 'only' require an immaterial solution and may also be very helpful in understanding systems, they contain very little guidance for the designer of physical products. These methods often include a latter phase that is called 'prototyping', but there is no specification as to how this prototyping should be done. Perhaps for this reason many who have worked with design thinking will recognise the Post-it Note as the typical material expression of the process (Hernández-Ramírez, 2018; Jen, 2017) and when physical prototypes are made, they tend to be small-scale mock-ups in model-making materials like cardboard or foam. These might be useful for discussing design principles, shapes and concepts, but provide no real information about the materials and the making.

Designing a physical product suitable for production following a design process that is based upon thinking is possible, but it requires a significant amount of prior practical experience with materials. When the material foundation from craft, which characterised the design education during the early part of 20th century, is absent, Design Thinking and similar immaterial design processes are fraught with the same flaw described by Henry Cole in the 1850s: a lack of contact with the materials for which the design were intended, which is inadequate when designing for production (Kriegel, 2007; Romans, 2005).

The rise of digital design during the latter part of the 20th century initially only reinforced the absence of the materials and making from the design process. Software programs were used to create 3D drawings, but these were primarily used for rendering to create visuals of design proposals. The academic Jeremy Myerson describes how digital design in many places became a convenient excuse to shut down expensive craft workshops and to lay off hands-on technicians (Myerson, 1997). However, when digital fabrication technology became accessible to designers and the public in general, this started to change with what has become known as the 'Maker Movement' (Anderson, 2012; Gershenfeld, 2005).

Whereas ample research in digital design and fabrication is available, research on the Maker Movement itself and the reasons why it started is relatively scarce. Nevertheless, it is possible to get an idea of the scale of the movement just by looking at the substantial number of Fab Labs that have opened, since the first one started in 2001. Fab Labs are mainly dedicated to digital fabrication and are part of a large open source network. The first Fab Lab was founded by Neil Gershenfeld, as a project out of the Center for Bits and Atoms at MIT and now there is an extensive network of Fab Labs spread out across the world (Fab Foundation, 2018). Even so, Fab Labs are just one type of maker space; there are many other varieties with different content and tooling, so the movement is not just based on digital fabrication, but rather on *design doing* with professionals and amateurs making everything from knitting workshops, to 3D printing, electronics, brewing beer and growing bacteria (Davies, 2017; Sheridan et al., 2014; Walter-Herrmann & Büching, 2014).

In relation to educational practice the Maker Movement is interesting because learning in these spaces is deeply embedded in the experience of making and doing. Educational psychologist Kimberly Sheridan and colleagues describe how the maker spaces value the process involved in making, through the activity of tinkering, in figuring things out and in playing with materials and tools (Sheridan et al., 2014). Many of the new Fab Labs are connected to educational institutions (Davies, 2017) and it would appear that at least some design schools are starting to shift their focus from a more conceptual and intellectual approach to include additional space for the tangible materials, skills and making. This aligns with the argument made by material scientist Mark Miodownik, who states that material scientists and engineers need to work with 'stuff', because Post-it Notes are simply not the right thinking tools for them (Miodownik, 2013).

Whether it is the Maker Movement that is provoking the awareness of materials and making at the design schools, it is hard to know; there is still only limited research on this. In the meantime, the indications of this are reduced to what can be observed in changes to the facilities and infrastructure of the schools – perhaps for the same reasons as motivated the opening of labs at Birmingham School of Art 120 years ago. Labs and workshops indicate that there is an awareness of the importance of doing, making and experimenting with physical matter and tools. It is perhaps too early to conclude if they will make a difference, as this entirely depends on whether the curricula will change accordingly.

2.5 The didactic challenge

As I have described, it would appear that over the last century the design profession has made an overall move from analogue to digital and from the hands to the head. And although this may be changing slowing, when trying to reintroduce materials into the design process and design education, it represents a didactic challenge. The reasons for this are described very accurately by the anthropologist Tim Ingold. Carrying a heritage from John Dewey's principles about learning by doing (Dewey, 1938), he identifies the value of 'knowing for oneself' and argues that in order to learn to practise something, e.g. designing a physical object, we must learn through 'participant observation'; we must study *with* and learn *from*. If we merely make a study *of* and learn *about* design we might be able to write an insightful piece about design, but we will not be able to design or make anything in practice (Ingold, 2013).

This clash between the formal or propositional and the informal or tacit with regards to knowledge is perhaps the most important quandary in educational practice within modern design education. Despite the fact that a design process based upon thinking does allow for a 'designerly way of knowing' and can be a good method for understanding systems, it is largely immaterial. It is based upon the principles from the Design Methods Movement in the 60s and 70s and thus is the work of excellent theorists, but theorists who for the most part had little or no experience with practice and making and perhaps therefore were unable to appreciate the value of the material foundation from craft.

Frayling describes how the distinction between formal knowledge and tacit knowledge – expressed from a social, as well as a technical perspective – was not clearly formulated at the beginning of the Design Method Movement, but only began with developments in the philosophy of knowledge (during the 1960s) and the sociology of science (during the 1970s) (Frayling, 2012). Perhaps for this reason the practical manipulation and handling of materials and tools, which will provide the designer with the embodied, technical and experiential information that is essential for designing a physical product, has gradually been pushed out. Naturally, design educations vary, but the general evolution over the last 50 years innately sparks scepticism that the status quo may not be easily compatible with a design process that reinstates working with materials for production.

The executed design method at Birmingham School of Art made the students work in the materials for which their designs were intended; the design process at the Bauhaus school was based upon creative material exploration while still including new technology and designing for industrial production; and Danish Modern was dominated by designers with backgrounds in crafts and who, consequently, had a substantial knowledge of materials and making. These three examples stand out when it comes to materials and making within the design process, because they carry a legacy from craft where the real material for production is present from the beginning of the design process and therefore contributes to informing, inspiring and restricting the design and product development.

References

Adamson, G. (2010). *The craft reader*. London, UK: Bloomsbury.

Adamson, G. (2013). *The invention of craft*. London, UK: Bloomsbury Academic.

Albers, A. (2018/1968). Interview with Anni Albers, 1968, July 5th. Archives of American art, Smithsonian Institution. Retrieved from https://www.aaa.si.edu/collections/interviews/oral-history-interview-anni-albers-12134#transcript.

Alexander, C. (1964). *Notes on the synthesis of form*. Harvard, MA: Harvard University Press.

Anderson, C. (2012). *Makers: the new industrial revolution*. New York, NY: Crown Business.

Archer, L. B. (1964). Systematic method for designers: Part seven. *Design, 188*, 56–59.

Ascher, B. E. (2015). The Bauhaus: case study experiments in education. *Architectural Design*, *85*(2), 30–33.

Bologna Declaration. (1999). The European higher education area. *Joint Declaration of the European Ministers of Education, 19*.

Broadbent, G. (1979). The development of design methods. *Design Methods and Theories*, *13*(1), 41–45.

Bullock, M., & Conrads, U. (1975). *Programs and manifestoes on 20th-century architecture*. Cambridge, MA: MIT Press.

Cole, H., & Redgrave, R. (1849). No title. *Design and Manufactures*, *1*(1).

Cross, N. (2006). *Designerly ways of knowing*. New York, NY: Springer.

Davies, S. R. (2017). *Hackerspaces: making the Maker Movement*. Hoboken, NJ: John Wiley & Sons.

Dewey, J. (2007/1938). *Experience and education* (2nd revised ed.). New York, NY: Simon & Schuster.

Donoso, S., Mirauda, P., & Jacob, R. (2018). Some ideological considerations in the Bauhaus for the development of didactic activities: the influence of the Montessori method, the modernism and the gothic. *Thinking Skills and Creativity*, *27*, 167–176.

Dormer, P. (1997). *The culture of craft*. Manchester, UK: Manchester University Press.

Droste, M. (2002). *Bauhaus, 1919–1933*. Cologne, Germany: Taschen.

Dyce, W. (1971/1843). *The drawing book of the Government School of Design*. London, UK: Chapman & Hall.

Fab Foundation. (2018). *FabAcademy*. Retrieved 03/01/2018, from http://www.fabfoundation.org.

Findeli, A. (1990). Moholy-Nagy's design pedagogy in Chicago (1937–46). *Design Issues*, *7*(1), 4–19. Retrieved from JSTOR database, http://www.jstor.org/stable/1511466.

Findeli, A. (2001). Rethinking design education for the 21st century: theoretical, methodological, and ethical discussion. *Design Issues*, *17*(1), 5–17.

Frayling, C. (2012). *On craftsmanship: towards a new Bauhaus*. London, UK: Oberon Books.

Gershenfeld, N. (2005). *Fab: personal fabrication, Fab Labs, and the factory in your computer*. New York, NY: Basic Books.

Hernández-Ramírez, R. (2018). Design thinking, bullshit, and what thinking about design can do about it. *Journal of Science and Technology of the Arts*, *10*(3), 2–45.

Hollingsworth, A. (2009). *Danish modern*. Layton, UT: Gibbs Smith.

Ingold, T. (2013). *Making: anthropology, archaeology, art and architecture*. Abingdon-on-Thames, UK: Taylor & Francis.

Jen, N. (2017). Design thinking is bullshit. Paper presented at the at 99U Conference, New York, New York.

Jones, J. C. (1966). Design methods reviewed. In S. A. Gregory (ed.), *The design method* (pp. 295–309). Boston: Springer.

Kriegel, L. (2007). *Grand designs: labor, empire, and the museum in Victorian culture*. Durham, NC: Duke University Press.

Lawson, B. (2006). *How designers think: the design process demystified*. Abingdon-on-Thames, UK: Routledge.

Library University of Glasgow. (2018). Images from *Journal of Design and Manufactures*, volume 1, London 1849. Retrieved from http://special.lib.gla.ac.uk/exhibns/month/aug2001.html.

Maldonado, T. (1960). New developments in industry and the training of designers. *Ulm*, *2*, 25–40.

Miodownik, M. (2013). The Institute of Making. *Materials Today*, *16*(12), 458–459.

Morris, W. (1888). The revival of handicraft. *Fortnightly*, *44*(263), 603–610.

Mussari, M. (2016). *Danish Modern: between art and design*. London, UK: Bloomsbury Publishing.

Muthesius, H. (1901). Kunst und maschine. *Dekorative Kunst*, 1901–1902.

Muthesius, S. (1998). Handwerk/kunsthandwerk. *Journal of Design History*, *11*(1), 85–95.

Myerson, J. (1997). Tornadoes, T-squares and technology: can computing be a craft? In P. Dormer (Ed.), *The culture of craft* (pp. 176–185). Manchester, UK: University of Manchester Press.

Rith, C., & Dubberly, H. (2007). Why Horst W. J. Rittel matters. *Design Issues*, *23*(1), 72–91.

Rittel, H. (1984). Second-generation design methods. In N. Cross (ed.), *Developments in design methodology* (pp. 317–327). Hoboken, NJ: John Wiley & Sons.

Romans, M., ed. (2005). *Histories of art and design education: collected essays.* Bristol, UK: Intellect Books.

Ruskin, J. (1971/1857). *The elements of drawing.* New York, NY: Dover Publications.

Schmiechen, J. A. (1990). Reconsidering the factory, art-labor, and the schools of design in nineteenth-century Britain. *Design Issues*, *6*(2), 58–69.

Schön, D. A. (1983). *The reflective practitioner: how professionals think in action.* Abingdon-on-Thames, UK: Routledge.

Semper, G. (2004/1860). *Style in the technical and tectonic arts, or, practical aesthetics.* Los Angeles, CA: Getty Publications.

Sheridan, K., Halverson, E. R., Litts, B., Brahms, L., Jacobs-Priebe, L., & Owens, T. (2014). Learning in the making: a comparative case study of three makerspaces. *Harvard Educational Review*, *84*(4), 505–531.

Simon, H. A. (1969). *The sciences of the artificial.* Cambridge, MA: MIT Press.

Swift, J. (2005). Birmingham and its art school: changing views 1800–1921. In M. Romans (Ed.), *Histories of art and design education: collected essays* (pp. 67–89). Bristol and Oregon: Intellect Books

Wallance, D. (1956). *Shaping America's products.* New York, NY: Reinhold.

Walter-Herrmann, J., & Büching, C. (2014). *FabLab: of machines, makers and inventors.* Bielefeld, Germany: Transcript Verlag.

Wright, T. (2014/1868). *The great unwashed.* Abingdon-on-Thames, UK: Routledge.

Yeomans, R. (1988). Basic design and the pedagogy of Richard Hamilton. *Journal of Art & Design Education*, *7*(2), 155–173.

Courtesy: Copenhagen School of Design and Technology, KEA.

3 The material dialogue in craft

Making things and tools has been part of human life since times immemorial, and as such craft is intrinsic to who we are and how we have built our civilisations. In order to understand the relationship between design, making and materials, it is necessary to examine their role within the history of craft. What makes the craft process particularly interesting in this context is the kind of dialogue with the material that takes place: from the outset, in craft there is a continuous dialogue between the maker, the material and the tool in the process of creating a physical object. It is relevant to study this dialogue for several reasons: in order to understand how the material can inform both the designer and the development of the product; not just to comprehend the value of physically handling materials and tools.

A study of recent literature on craft reveals how it can include art, design, industry and even rituals and makes it clear that it is indeed impossible to describe craft as one 'thing' (Adamson, 2010, 2013, 2017; Dormer, 1997; Frayling, 2012; Pye, 1968). This is also why the discussion in this chapter by no means encompasses all elements from craft that are relevant to design, but simply examines a few fragments that are of particular interest in order to understand the meaning of designing with materials, why craft was separated from design and the potential for bringing it back.

3.1 The separation of mind and body

The extensive research on the history of craft by the historian and curator Glenn Adamson shows that not only have many different definitions of craft been proposed, but the last 200 years of craft have not been an unobtrusive journey (Adamson, 2010, 2013, 2017). Today, the term 'craft' tends to be used for small-scale handmade productions and is both identified with expert skill but also suffers from the stigma of amateurism and hobbyism. Perhaps the frictions in the history of craft are reflected in its present perceived value, which appears rather variable. This discrepancy can be exemplified by expensive shoes or clothing labelled 'handmade': something that is typically perceived as a stamp of quality, indicating diligence and authenticity. However, the garment will still most likely carry the name of the designer, whereas the craftsperson, who is the actual maker and may even have played an important role in the development of the design, typically remains anonymous. This offers a strong indicator of his or her hierarchical position in comparison to that of the designer, and, as will be outlined below, might also imply that imagining the object is valued more than making it.

The act of craft implies skill and knowledge about a material or a group of materials. In German, the term for craft is *Handwerk* (and likewise in some related languages), which literally translates into 'work of the hand'. During the Industrial Revolution a noticeable separation began between the artistic intellectuals imagining the object and the maker of the object itself. Plenty has been written about this transition and characteristically this literature is focused on the loss suffered by the manual worker, who without the mental activity and creative imagination needed for the design seemingly was reduced to performing menial labour that was directed by others. Karl Marx was perhaps the strongest early opponent of this separation because he believed factory work to be dehumanising and pointed out how 'In handicraft and manufacture, the workman makes use of a tool, in the factory, the machine makes use of him' (Marx, 1867).

The Marxist Harry Braverman described much later how the

> separation of mental work from manual work reduces, at any given level of production, the need for workers engaged directly in production, since it divests them of time-consuming mental functions and assigns these functions elsewhere ... The production units operate like a hand, watched, corrected and controlled by a distant brain.
>
> (Braverman, 1974, p. 86)

This describes the idea of a separation between the mind and the work of the hand. Repetitive manual work *can* naturally be both a physical and mental burden and it does not produce the satisfaction of imagining something new that does not yet exist and then subsequently making it. This kind of repetitive manual work does not hold what the craftsman and former professor at the Royal College of Art David Pye calls 'free workmanship of risk' as opposed to 'workmanship of certainty' (Pye, 1968).

Descriptions of the division of labour during and after the Industrial Revolution often paint a picture of the craftsperson as incapacitated in the transition; as if there were no longer any need for free, intelligent, creative or complex thinking in practical manual work (Marx, 1867; Morris, 1888). Adamson describes how the history of craft is in some respects more diverse: that craft in many ways pushed forward technology where some craftspeople became specialists who indeed possessed the mental resources described above. Furthermore, he also suggests that the division of labour as well as the exploitation of the labour force long predated the Industrial Revolution (Adamson, 2013). Understanding the differences in the cognitive processes between machine-made products and (truly) handmade craft is interesting to explore, because it may hold some lessons as to how one might transfer the material dialogue from craft into a context of digital design and manufacturing, which are likely to be part of a contemporary design process.

Another important aspect is the narrative of the craftsperson suffering from the division of labour. By no means do I wish to question records of poor working conditions in factories during the Industrial Revolution. However, what is interesting within the context of this book is the narrative of the division which appears to be built upon the ideas of dualism and a split between body and mind: where the creative intellect is separated from the hands, the liberal arts from the mechanical arts, theory from practice, thinking from making, etc. The division between mind and body is a position that can be traced a long way back. Perhaps the most famous quote from the philosopher René Descartes is revealing and apt at this juncture: 'I think therefore I am' (Descartes, 1637) and even in antiquity Plato proclaimed that 'the eye, ears and the whole body a disturbing element, hindering the soul from the acquisition of knowledge' (Plato, 2013) [approx. 360 BC].

However, the people writing about the division of labour tend to be philosophers, sociologists, historians, scientists and economists (with metalworker Thomas Wright a notable exception (Wright, 1868)). In general, such people are not makers themselves and have not experienced the intellectual and creative aspects involved, or the value and information derived from manipulating materials and working manually. They comprehend and represent the intellectual side of the separation, the side of which the craftsperson was supposedly deprived. Consequently, one could claim that it appears to be a one-sided story.

According to the historian Paul Greenhalgh, the discourse of separation based upon a Cartesian duality was only reinforced during the 20th century, when fine arts and craft were separated as domains. The result was that craft and making became considered a 'non-intellectual' activity that was based on manual skill and therefore it lost the status that it had regained with the philosophy developed by the Arts and Crafts Movement. He describes that as early as the 1920s there was a shift away from understanding creative practice as an inseparable part of the physical process of making, where cognitive and manual activities were effectively considered the same, to craft being identified as non-intellectual. Despite the consequences and changes to design provoked by divorcing craft, it did not affect design's status negatively, but rather the ultimate outcome for design was losing the making and the cognition that goes with it (Greenhalgh, 1997). Thus, considering that the design profession over the last 100 years has increasingly been defined by thinking, as designers we should at least consider if we might have lost something of value in the process of separating our profession from that of craft and making.

3.2 Cognition in craft: the bond between body and mind

The aim of this section is to understand the value of practical expertise in craft, more specifically, to highlight how essential the coupling of body and mind is to the successful manipulation of materials and tools within the process of designing and making. Some descriptions of the composition of craft are focused not just on technical and practical skills, but also on the bodily kinaesthetic intelligence involved in the process of making. The process of craft involves all senses and provides the maker with feedback from interactions with the materials, the tools and the environment. This not only leads to a more intimate knowledge of the materials, but also appears to affect one's state of mind.

For ceramicist Hal Riegger, working with primitive ways of making pottery induced clear logical thinking. He describes it as being taken back to the uncluttered thinking of children, which he describes as not only refreshing but also good training for mental processes (Riegger, 1972). The mental aspects also seemed to be important to Otto Salomon, who invented the Scandinavian 'Slöjd' system in the 1870s. He introduced the teaching of woodworking as a part of the curriculum in comprehensive schools. In this context, the interesting fact is that his motivation was not just to train manual labourers, but also to bring forward qualities in the pupils, such as pleasure in bodily labour, discipline, mental strength and the power of observation and execution by hand (Salomon, Nordendahl & Johansson, 1894).

The act of craft is frequently described as producing a sense of joy and satisfaction (Buechley & Perner-Wilson, 2012; Richards, 2011; Schumacher, 1973). Perhaps for this reason, craft is at times used during and as therapy. There is some evidence that craft can help reduce stress and bring about change in emotions, bodily and sensory experiences, patterns of thought, actions and behaviour (Goodman & Rosner, 2011; Pöllänen, 2009; 2015). The reason for addressing this is not to focus on the potential benefits of craft for improving mental health, but to draw attention to the fact that in a craft process there appears to be a significant connection between the embodied cognitive experience of working with the tools and materials and the mind. There is no need to justify this to a craftsperson, because he or she will know this intuitively.

George Sturt, who in the 1920s wrote about rural craft, asserted that it is only through direct, physical experience that one can understand workmanship or even raw materials: 'my own eyes know because my hands have felt, but I cannot teach an outsider, the difference between ash that is "touch as whipcord" and ash that is "frow as carrot", or "doaty" or "biscuit" ' (Sturt, 1923, p. 24). However, to researchers who have not been actively involved in making and working with materials the importance of the bodily engagement in the process of craft might require further explanation.

The theory behind embodied cognition argues that it is not just shaped by the brain, but by aspects of our entire body and its interactions with the environment (Shapiro, 2010). In his book *The ecology of the brain* the German psychiatrist and philosopher Thomas Fuchs breaks with the idea that the brain steers us like marionettes from some hidden 'headquarters', and describes how the brain effectively should be understood as an organ only animated in connection with our senses, nerves and muscles, with the internal organs, our skin, our environment, and in relation to other human beings (Fuchs, 2018). In relation to design and the design process this is very interesting, because it provides an explanation of the way our senses and the motor system are involved in our understanding and working with materials and tools. It makes it possible to comprehend the importance of the information received through our hands and senses in the process of making a physical object.

It is by no means entirely clear how the mind works and this is reflected, among other places, in the controversies between standard cognitive science and embodied cognition; however, most cognitive scientists appear to agree that many cognitive processes happen unconsciously (Shapiro, 2010). With practice, tasks that initially were complex and required one's full attention become automatised (Wilson, 2002). Andy Clark calls this automaticity 'offline' cognition as opposed to 'online' cognition, which can be described as intentional actions or time-constrained real-world interactions (Clark, 1998). There is a limit to human 'online' cognition and thus the body is forced to handle as much as possible 'offline'. Activities that initially evolved for perception and action 'online' will, once embedded, run 'offline'. 'Offline' cognition is of great importance in sensorimotor co-ordination skills needed in an endless number of situations, such as cycling, playing piano, in object manipulation or the handling of tools. 'Offline' cognition is bodily based, but should not be considered inferior. On the contrary, as will be exemplified below, running 'offline' appears to support higher-level cognitive abilities, which confirms the notion that humans may 'think' faster with the hands (body), than with the brain (Clark, 1998; Kolbeinsson & Lindblom, 2015).

These two kinds of cognition are not to be understood as completely separate, but rather as parallel modes that can be switched between or used simultaneously. Functioning 'offline' perhaps clarifies how it is indeed possible for a musician to play complicated sheet music, at a speed where there is no time to look at each written note separately and to consciously make the decision to play a specific tune, or why football players will explain that they missed a shot because they stopped to think, or how it is possible for us react to dangerous situations in a split second and how we can perform an activity that requires advanced sensorimotor coordination, such as cycling or running, and yet at the same time be thinking about something else. Equally, it might at least partly explain the expert cognition in craft as a predominantly 'offline' embodied dialogue with a material and a tool.

Ari Kolbeinsson and Jessica Lindblom describe that,

> much of skilled tool use and assembly is anticipatory, and that the expert's skill comes from the ability to control their activity with sufficient spare capacity to cope with future demands and to respond to the changing circumstances in which they are acting to effect changes in the objects being worked on, i.e. the 'cognitive space' in which subsequent decisions are made.
>
> (2015, p. 5189)

Embodied cognition and 'running offline' provide an explanation for the importance of engaging physically with a material in order to fully understand its potential. However, the question remains of how to obtain the same level of cognition in digital design and manufacturing, where there is no apparent need for manual handling of the materials.

3.3 Analogue versus digital

As described in the previous sections, craft has a very close connection with the hands. However, craft does make use of tools. Adamson describes how craft is almost always a matter of triangulation between maker, tool and material. In this process the naked hand might be considered a tool, but he also points out that there is no obvious reason why any particular type of tool should be considered ineligible for this relation (Adamson, 2010). The 20th-century craft discourse tends to see the machine as a threat to craft and the cause of a decline in traditional skills. Craft was generally understood as being in opposition to industry, technology and modernity. However, architects like Hermann Muthesius and Frank Lloyd Wright did not believe that the machine was in opposition to craft, but rather they saw it as an aid to creativity. In the beginning of the 20th century Muthesius argued that art should take the machine-made under its wing in order to survive (H. Muthesius, 1901). The machine was to be the architect's tool, whether they appreciated it or not, and it was the architect's job to master the machine (Wright, 1975).

In digital design and manufacturing processes today, machines and digitalisation facilitate the production of very complex forms by the means of advanced production methods. In 1998, the academic Malcolm McCullough wrote about how he believed that digitalisation could potentially restore the importance of craft and that digital design and manufacturing would make small-scale production competitive with large-scale manufactures (McCullough, 1998). In many ways his ideas are proving true and can be observed in the Maker Movement that is flourishing today through Fab Labs, tech shops and shared workshops. These many variations of making represent a great diversity in techniques, processes, materials and form – something that Pye believed to be unique to craft and the main reason for perpetuating craftsmanship. He claimed that this freedom of diversity was usually impossible for artists and designers, because artists were constrained by the culture of contemporary art and designers were controlled by the demands of industry and mass production (Pye, 1968).

The Maker Movement is without doubt bringing back many elements from craft. However, going back to the foundation of traditional crafts and the embodied cognition that takes place in the dialogue between the maker, the material and the tool raises the question of what happens to this dialogue when the process is largely digitalised, and consequently without tangible contact with the material. The sensorimotor interaction in this case primarily becomes the touch of the keyboard and as a result the main sensory stimulus the designer receives is visual. Fuchs describes how instruments can be integrated as an extension of the body and how the limits of the subjective, lived body do not always correlate exactly with the limits of the objective body. An example of this could be a blind man with a walking cane, who does not feel the resistance of the surface being grasped for in his hand, but rather at the cane's tip, or a trained driver who notices the quality of the street coating under the tyres of his car (Fuchs, 2018).

3.4 Matter versus form

Traditional master craftspeople would search for the limits of materials, but the material would always influence and restrict the form. Yet, with many new materials and manufacturing processes invented during the late 19th century and the beginning of the 20th, these constraints began to change, at least to some degree. Materials such as papier-mâché, rubber and early types of synthetic plastics broke with the material categorisation from traditional craft, had exceptional properties

(a)

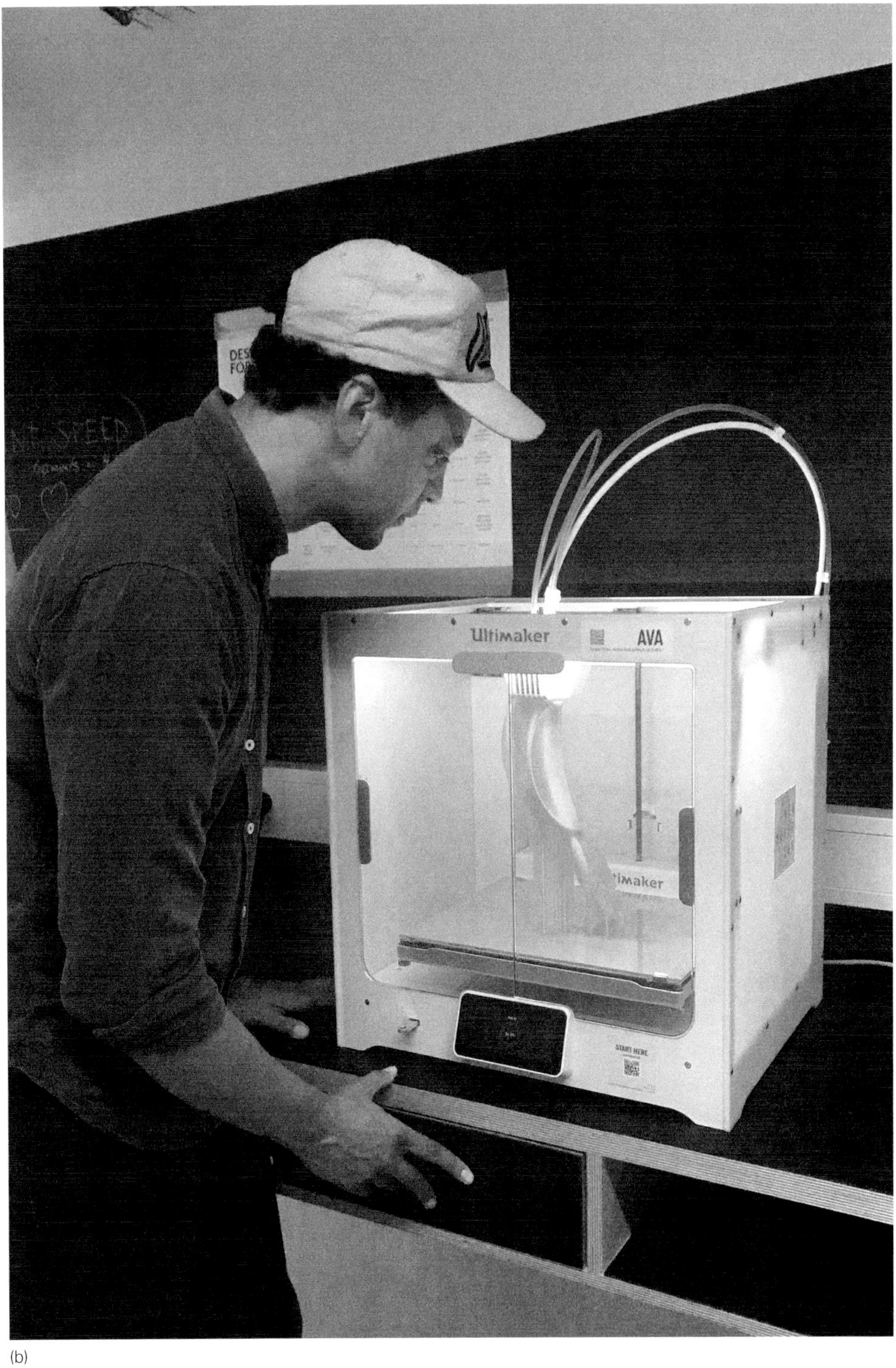

(b)

Figures 3.1 a and 3.1b **Analogue versus digital manufacturing**
Very few types of craft are in fact executed with the bare hands. E.g. glassblowing, as can be seen in Figure 3.1a, is a process in which the craftsperson (for obvious reasons) never touches the material with the naked hands, but which still requires great skill and sensorimotor interaction with the material. During this process, the level of fluidity of the molten glass is felt through the resistance in the lungs and on the lips touching the steel pipe when blowing, the centring and the weight of the glass are felt at the tip of the pipe when the glass is being turned. However, in the case of automated tools such as the 3D printer in Figure 3.1b, laser cutters or CNC milling machines the user is prevented from steering the tool through fine motor control and thus largely loses the experiential contact with the material. These considerations are important when reintroducing the material dialogue from craft into a contemporary design process that also involves digital design and fabrication.

and were very easily mouldable. Semper described how form had always arisen through the logic of craft – the interaction of materials, tools and function, rather than through conscious imposition. Now, those various types of resistance were banished (Adamson, 2013; Semper, 1860).

The art historian Henri Focillon was right, from the perspective of natural sciences, when claiming that 'form does not behave as some superior principle modelling a passive mass, for it is plainly observable how matter imposes its own form upon form' (Focillon, 1942). But even so, the significant workability of the new materials allowed for considerably more liberty when imagining form on its own. Paradoxically, the invention of new materials in many respects reinforced the designer's separation from the control of materials and making, and it likely also contributed to the division between form and matter, which frequently characterises the design process today.

In his book *The eyes of the skin*, the architect Juhani Pallasmaa describes how the dominance of vision over the other senses provokes a distance to, and separation from, the surroundings, and like the cognitive theories presented earlier in this chapter, he too emphasises that it takes all of the senses to truly grasp materiality (Pallasmaa, 2012). Vision enables us to easily comprehend shapes and contours and perhaps, therefore, naturally makes us focus on form. But whereas we can appreciate forms on a bigger scale, the human eye is limited; with the naked eye we cannot see the protein that constitutes silk, nor can we appreciate the molecular structure that makes wood fibres flexible. Nature does not separate materials from form, nor does science. In fact, one could go as far as to say that it is the material that drives the form in any living organism, and looking into an optical microscope will make it evident that the material already has a form.

It would appear that the design process has lost the material dialogue it inherited from traditional crafts. There may be several reasons for this loss: Perhaps it is a result of being culturally and physiologically dominated by our vision. Perhaps it is due to the complexity and overwhelming amount of new materials that have been invented in recent history. Or perhaps, it has been caused by the last 60 years of scientific and intellectual tendencies in design education. Nevertheless, the result is a condition in which matter and form are largely separated in the process of creating a physical object. It is a design process where the material, for which the design is intended, if it appears at all, only does so at the very end of the process. This reduces product designers to inventors of form, ideas and concepts, who do not understand the matter that constitutes the physical realm of the world that they are designing for. It leaves them without control of their own design, without the ability to design for the criteria of sustainability and it deprives them of the capability to explore new types of materials that might represent valuable and sustainable alternatives.

3.5 Reinstating materials and making

In 2012, Christopher Frayling asked: 'Today, with the decline of laboratory work in school science, and the transformation of design into a briefing process, "making" is at its lowest ebb since before the Arts and Crafts Movement. Can it come back?' (Frayling, 2012). The Maker Movement suggests that this may be coming back and equally there are some indications that the same could be the case with materials. Materials libraries, like the Material Connexion (www.materialconnexion.com) represented in different places around the world, the distinctive MaterFad materials collection (www.materfad.com) that started in Barcelona or the open source initiative Materiom (www.materiom.org), which provides recipes for different biomaterials, would indicate that there is a growing interest in new materials.

The same is traceable within both design practice and research. The work of designers such as Suzanne Lee, Maurizio Montalti, Maartje Dros, Eric Klarenbeek and Jonas Edvard is solidly founded on material exploration and experimentation, and in their designs they push forward the use of different new biomaterials and processes. Lee and Montalti have a strong connection with natural science and biotech, Dros and Klarenbeek with technology and art, and Edvard has roots in the Danish design tradition (Dros, Klarenbeek and Edvard share insights from their design processes in interviews included in Chapter 4). Pirjo Kääriäinen et al. describe how experimental material projects and material design are becoming increasingly popular among designers and in this way expanding the design profession. They point out how designers can use their imagination and skills to utilise waste streams or to create new materials from unexpected sources (Kääriäinen, Tervinen, Vuorinen & Riutta, 2020).

Equally there are several research groups exploring new materials and how to design with materials, encompassing a wide variety of approaches and content from the intricate material-based projects related to biology and additive manufacturing led by the architect Neri Oxman from the Media Lab at MIT, the broad portfolio of projects related to material technology and making directed by the material scientist Mark Miodownik and artist Zoe Laughlin at the Institute of Making at UCL, to the work on material-driven design led by Elvin Karana at Delft University.

Material-driven design, or material-based design as it is called by other researchers, is a design method or approach where the design is inspired by the exploration of a material (Hansen, 2010; Karana, Barati, Rognoli & Zeeuw van der Laan, 2015; Lussenburg, van der Velden, Doubrovski, Geraedts & Karana, 2014; Oxman,

2010; Van Bezooyen, 2013). In the book *Materials experience*, the designer Aart van Bezooyen states that 'material driven design' is fundamentally like turning the 'normal' design process upside down: instead of starting with designing the form and leaving the material selection to the very end, one actually starts with the material (Van Bezooyen, 2013).

This observation is noteworthy because it reflects that the present more form-focused and immaterial design process could be seen as an established norm. Consequently, although materials and making may be on the way back, it could prove difficult to change the role of materials in the design process. Nevertheless, as documented earlier in this chapter, the design process has never been one thing, and since the first design schools opened 200 years ago, materials have moved in and out of this process. Consequently, the status quo can hardly be seen as a final truth and there is no reason why materials and making cannot find their way back into the design process.

As already described, the Bauhaus followed design principles and processes in which materials were central, while at the same time industrial production and technology were embraced by Bauhaus designers. The Bauhaus curriculum stands out as exemplary of a design education with a strong focus on material exploration (Itten, 1975; Karana et al., 2015; Rognoli, 2004; Wick & Grawe, 2000). An updated version of the Bauhaus foundation course and elements from the traditional craftsmanship-based design educations could be very relevant within a modern design education context. Nevertheless, contemplating today's vastly augmented quantity and complexity of materials and new advanced production processes, it is also evident that the solution is not to re-divide the design profession into traditional material workshop categories like wood, glass or ceramics; nor can we ignore theory or new technology. In the near future that designers will have to relate to, the most readily available materials are likely to be recycled materials and biomass by-products.

Furthermore, new materials are emerging from fields that traditionally have had very little to do with design: for example a broad range of professions, industries and researchers, including the medical industry, are using tissue engineering to develop new products and synthetic biology to grow new degradable biological materials in laboratories. Miodownik highlights how this complexity and the amount of materials are likely to increase (Miodownik, 2015). Still, with Sarah Wilkes and colleagues, he also describes how designers increasingly have the opportunity to influence the development of materials as they emerge from the laboratory, if they are able to communicate with material scientists (Wilkes et al., 2016).

Thus, if materials are to be central once again in design education and the design process, it is clear that the changes required will not simply be achieved by introducing a short course on materials or providing sufficient books on material technology within the existing framework. To reintroduce the material dialogue from craft into a contemporary design process fundamentally changes the way we design things and, thus, inevitably it affects both the constituents of the design process and the way we teach design. I will address the details, the meaning and the effect of this in the following chapters.

References

Adamson, G. (2010). *The craft reader*. London, UK: Bloomsbury.

Adamson, G. (2013). *The invention of craft*. London, UK: Bloomsbury Academic.

Adamson, G. (2017). *Thinking through craft* (5th ed.). London, UK: Bloomsbury.

Braverman, H. (1974). *Labor and monopoly capital: the degradation of work in the twentieth century* (25th ed.). New York, NY: Monthly Review Press.

Buechley, L., & Perner-Wilson, H. (2012). Crafting technology: reimagining the processes, materials, and cultures of electronics. *ACM Transactions on Computer–Human Interaction (TOCHI)*, *19*(3), 21.

Clark, A. (1998). *Being there: putting brain, body, and world together again*. Cambridge, MA: MIT Press.

Descartes, R. (1968/1637). *Discourse on method and the meditations*. London, UK: Penguin.

Dormer, P. (1997). *The culture of craft*. Manchester, UK: Manchester University Press.

Focillon, H. (1942). *The life of forms in art* (G. Kubler Trans.). New Haven, CT: Yale University Press.

Frayling, C. (2012). *On craftsmanship: towards a new Bauhaus*. London, UK: Oberon Books.

Fuchs, T. (2018). *Ecology of the brain*. Oxford, UK: Oxford University Press.

Goodman, E., & Rosner, D. (2011). From garments to gardens: negotiating material relationships online and 'by hand'. In *Proceedings of the SIGCHI Conference on Human Factors in Computing Systems* (pp. 2257–2266). Vancouver, Canada.

Greenhalgh, P. (1997). The history of craft. In P. Dormer (Ed.), *The culture of craft* (p. 20). Manchester, UK: Manchester University Press.

Hansen, F. T. (2010). Material-driven 3D digital form giving, experimental use and integration of digital media in the field of ceramics. Doctoral dissertation, the Danish Design School.

Itten, J. (1975). *Design and form: the basic course at the Bauhaus and later*. Hoboken, NJ: John Wiley & Sons.

Kääriäinen, P., Tervinen, L., Vuorinen, T., & Riutta, N. (2020). *The CHEMARTS cookbook*. Helsinki, Finland: Aalto University.

Karana, E., Barati, B., Rognoli, V., & Zeeuw van der Laan, A. (2015). Material driven design (MDD): a method to design for material experiences. *International Journal of Design*, *9*(2), 35–54.

Kolbeinsson, A., & Lindblom, J. (2015). Mind the body: how embodied cognition matters in manufacturing. *Procedia Manufacturing*, *3*, 5184–5191.

Lussenburg, K., Van der Velden, N., Doubrovski, E., Geraedts, J., & Karana, E. (2014). Designing with 3D printed textiles: a case study of material driven design. Paper presented at iCAT 2014: 5th International Conference on Additive Technologies, Vienna, Austria.

Marx, K. (1867). In Engels F. (Ed.), *Capital: volume one*. British Colombia, Canada: Modern Babarian Press.

McCullough, M. (1998). *Abstracting craft: the practiced digital hand*. Cambridge, MA: MIT Press.

Miodownik, M. (2015). Materials for the 21st century: what will we dream up next? *MRS Bulletin*, *40*(12), 1188–1197.

Morris, W. (1888). The revival of handicraft. *Fortnightly, 44*(263), 603–610.

Muthesius, H. (1901). Kunst und maschine. *Dekorative Kunst, 1901–1902*.

Oxman, N. (2010). *Material-based design computation*. Unpublished doctoral dissertation, Massachusetts Institute of Technology, Cambridge, MA (665764026).

Pallasmaa, J. (2012). *The eyes of the skin: architecture and the senses*. Chichester, UK: John Wiley & Sons.

Plato. (2013). *Phaedo. The last hours of Socrates* (Benjamin Jowett Trans.). Retrieved from, http://www.gutenberg.org/files/1658/1658-h/1658-h.htm.

Pöllänen, S. (2009). Craft as context in therapeutic change. *The Indian Journal of Occupational Therapy*, *41*(2), 43–47.

Pöllänen, S. H. (2015). Crafts as leisure-based coping: craft makers' descriptions of their stress-reducing activity. *Occupational Therapy in Mental Health*, *31*(2), 83–100.

Pye, D. (1968). *The nature and art of workmanship*. Cambridge, UK: Cambridge University Press.

Richards, M. C. (2011). *Centering in pottery, poetry, and the person*. Middletown, CT: Wesleyan University Press.

Riegger, H. (1972). *Primitive pottery*. New York, NY: Van Nostrand Reinhold.

Rognoli, V. L. (2004). How, what, and where is it possible to learn to design materials. Paper presented at the 7th International Conference on Engineering and Product Design Education, Delft, Netherlands.

Salomon, O., Nordendahl, C., & Johansson, A. (1894). *The teacher's handbook of slöjd*. London, UK: G. Philip & Son.

Schumacher, E. F. (2011/1973). *Small is beautiful: a study of economics as if people mattered*. New York, NY: Random House.

Semper, G. (2004/1860). *Style in the technical and tectonic arts, or, practical aesthetics*. Los Angeles, CA: Getty Publications.

Shapiro, L. (2010). *Embodied cognition*. Abingdon-on-Thames, UK: Routledge.

Sturt, G. (1963/1923). *The wheelwright's shop*. Cambridge, UK: Cambridge University Press.

Van Bezooyen, A. (2013). Materials driven design. In E. Karana, O. Pedgley & V. Rognoli (Eds.), *Materials experience, fundamentals of materials and design* (pp. 277–286). Oxford, UK: Butterworth-Heinemann.

Wick, R. K., & Grawe, G. D. (2000). *Teaching at the Bauhaus*. Berlin, Germany: Hatje Cantz Publisher.

Wilkes, S., Wongsriruksa, S., Howes, P., Gamester, R., Witchel, H., Conreen, M., et al. (2016). Design tools for interdisciplinary translation of material experiences. *Materials & Design*, *90*, 1228–1237.

Wilson, M. (2002). Six views of embodied cognition. *Psychonomic Bulletin & Review*, *9*(4), 625–636.
Wright, F. L. (1975). *In the cause of architecture*. New York, NY: Architectural Record.
Wright, T. (2014/1868). *The great unwashed*. Abingdon-on-Thames, UK: Routledge.

4 Reintroducing materials into a contemporary design process

Reintroducing what can be seen as a material dialogue from craft into a contemporary design process naturally calls for an understanding of the role of materials in a craft process. Nevertheless, it is important to recall that the intention with this book is not to suggest that we as designers 'return' to a pure craft process or that we should make all design education material-specific, as it still is with some fields like ceramics or glass. The ambition is to embrace the material dialogue that exists within a craft process, where information is retrieved by interacting with the physical material intended for the actual design, and transfer this approach to materials into a contemporary design process.

Discussing dialogues as part of the design process is not something new. Donald Schön describes designing as having a reflective conversation with the materials of a situation (Schön, 1983, 1992). The act of designing could well be defined as a series of dialogues: with users, technology, manufacturers, drawing, fashion, ergonomics, tradition, etc. However, the dialogue in focus here is the designer's dialogue with *real* materials. That is, the materials from which the final design is expected to be produced and not 'just' representational materials typically used for building models. Because although these types of materials may be very valid for informing the designer of proportions and form, they are merely representational. Therefore, before describing a contemporary design process that includes working with materials, it is useful to outline some fundamental aspects of a material dialogue as it appears in its 'pure' form in craft and understand how these can be transferred and translated into a contemporary design process.

4.1 From craft to design

In simple terms, the material dialogue found in craft can be described as the dialogue that takes place between a craftsperson, a material and a tool. The dialogue can appear in many variations depending upon the expertise of the craftsman, the quality of the material and the type of tool. For obvious reasons the craftsperson is the only part in the dialogue capable of making conscious decisions, but the material and the tool will influence the dialogue as well. The material always has a resistance, but it might also have unexpected properties and even flaws. These will challenge and inspire, but they will also force the craftsperson to adapt to the material. Likewise, the type of tools used and one's ability to control these tools will affect decisions in the design process and the appearance of the final result.

It is through this dialogue that the craftsperson achieves an embodied understanding of the material and gains the practical expertise and fine motor control required to manipulate the material and tools skilfully. It provides the craftsperson with a type of experiential knowledge that makes it possible to imagine how a material might behave or look in specific circumstances – even when the material is not at hand. Practising and repetition lead to expertise, and a master craftsperson who has produced several objects of the same kind will have gained a proficiency in their production and a familiarity with the materials involved.

An exemplary description from 1895 of what it means to gain expertise through a material dialogue can be found below in the written appreciation of traditional craftsmanship by the art dealer Siegfried Bing:

> First, try to enrich the arsenal of usable materials with every element in nature, down to the lowliest, which the blinders of old habits have until now ignored. To manipulate these materials, master every known process and method, in their most diverse applications. Then, after we have learned and analysed everything, acquired every secret technique, every trick of the trade as taught by the experience of centuries ... then, completely forget the way these have been used in the past, banish from memory any lingering obsession with inherited forms; in a word, place old and tried knowledge in the service of an entirely new spirit, with no guidelines other than those of intuitive taste and natural laws of logic.
>
> (Bing, 2010, p. 101)

With expertise in manipulating a material comes automaticity, the 'offline' cognition that I described in Chapter 3, and also what is often described as tacit knowledge. These can be hard to explain, perhaps because there isn't necessarily any need for words in the process and therefore the craftsperson may not have the words to explain the complexity of the embodied knowledge and mental activity involved. Glenn Adamson describes this as the internal coherence of a craftsperson, who will continuously adjust the inputs according to what is often described as subconscious, instinctive or experiential knowledge (Adamson, 2010). To the inexperienced observer it makes watching the process of a craftsperson throwing a vase in clay, or turning the wooden leg of a table, mesmerising. However, the lack of words might also mislead the observer into believing that what takes place is merely of a practical and vocational character.

Adamson points out that craft always entails an encounter with the specific properties of a material or materials (Adamson, 2017). The craftsperson is indeed trained to know about the material and how to manipulate it. Therefore, he or she will relate to the material even when imagining new designs and the material is not immediately at hand. With respect to this, perhaps the craft process could in itself be described as a material dialogue. A craftsperson's in-depth knowledge of a material and their inherent skills in making inevitably result in a different material dialogue than that which a designer may experience.

A common difference between the designer and the craftsperson is that the latter is also the maker of the final product. The craftsperson is likely to have a much higher skill level, an extensive knowledge about a specific category of materials and has been trained in how to manipulate them. Thus, whereas the craftsperson is likely to be the maker of the final product, the designer will typically be expected to produce a prototype, at most. For the craftsperson this requires a level of skill and control over materials and tools, which demands extensive practice – a level that is likely to be attained only if the activity is limited to one category of materials and its associated tools and specific techniques. The designer is not expected to be the maker of the final product, but, on the other hand, is expected to design products for a wide variety of materials. Therefore, the designer is unlikely to match the level of skills possessed by a craftsperson, and while the quality of the material dialogue may not be equal to that of a craftsperson, the designer will still receive the same type of experiential knowledge through the manipulation of a material.

The traditional crafts are divided into material categories, like ceramics, glass, wood, metals, etc. These categories essentially define the tooling and build on a heritage of particular techniques and tradition. But although most materials might lend themselves to certain tools, like wood to woodworking tools, a more open and explorative material dialogue can involve the use of tooling from other material groups, or even require building special tools to suit the material and process.

In my research I observed that design students have a more open approach to tooling in their experimentation with materials. This variety is particularly evident when the students are working with more unusual materials that do not fit into any classic craft workshop category; e.g. biomass waste from the food industry. Considering the rudimental making skills of some design students, the uninhibited approach to tools could be a result of not having received any technical training before starting the material exploration. This disadvantage can often result in rather inept prototypes, but potentially also has the advantage of opening up for creative solutions unrestricted by tradition.

Interestingly, at the Bauhaus School an initially naïve approach to material exploration was encouraged as something positive. Johannes Itten's foundation course at the Bauhaus would deliberately let the students explore materials without introducing them to specific techniques first (Itten, 1975). In an interview, Anni Albers, who was initially a student and later became a teacher at the Bauhaus, emphasised how in the absence of established techniques students would invent new ways of manipulating the materials and making things (Albers, 1968).

Even though a craftsperson would seem to be better trained for a material dialogue aimed at a finished functional product, perhaps the lack of schooling in established techniques does to some degree enable a more creative exploration of how one can manipulate a material. Nevertheless, in the Bauhaus curriculum this free exploration was mainly dedicated to the foundation course. After this the students would continue their schooling in workshops with more standard equipment and conventional facilities, and, as will be elaborated in Chapter 5, there can be little doubt that introducing a material dialogue into a design process demands skills in making.

For a designer with reasonable making skills the main difference between a material dialogue in craft and in design are the reasons for using it: A material dialogue may be enjoyable and inspiring to a craftsperson, but it is also likely the means to an end that will result in a finished product. To the designer, the material dialogue is a way to learn about the material reality of the design. However, it is not to be understood purely as a practical or technical activity; as I will describe below, it is in fact closely connected to creativity and may even contribute to reflective thinking.

4.1.1 Analysis and synthesis

A design process can be described as the journey the designer takes from framing an idea to presenting a solution. It may take various forms, employ several design methods and be more or less systematic. Design processes are usually conceptualised in terms of an iterative sequence of ideation, or finding a problem, drafting ideas, creating a product, reflecting, and revising (Cross, 2006). This complexity means that the design process has no single recipe. Even when a design brief initially introduces a series of constraints, the designer may introduce additional constraints from domain knowledge, or during the exploration of a particular solution (Ullman, Dietterich & Stauffer, 1988). Yet, there are different activities that should be present in the process. These involve either *analysis* or *synthesis*, or both.

Analysis is concerned with problem definition. The designers must understand and determine the framework in which the solution has to be designed: they must understand the systems involved, like the system for sustainability introduced in Chapter 1, or a specific culture, market or user group, in order to understand all of the requirements. *Synthesis* is concerned with problem solution, with proposing and testing that which does not yet exist. During the Design Methods Movement in the 1960s there was a belief that analysis and synthesis should be divided into two distinct phases. However, as many researchers and designers have since pointed out, in actual practice a design process is not divided in this way (Buchanan, 1995). Consequently, both analysis and synthesis should be present within the design process, not separated into linear phases but rather intertwined, so that they inform one another.

In the 1950s, designers had a tendency to solve problems by synthesis rather than analysis. They would focus on creating a solution for the physical product without necessarily fully understanding the problem they were trying to solve or the user they were designing for (Lawson, 2006). This appears to have changed, perhaps as a result of today's strong focus on Design Thinking or due to a more solid theoretical foundation. However, in a process of designing what does not exist yet, synthesis must at some point in the process take precedence. For this to take place, the designer needs imagination and tools like drawing to support and represent new ideas, but the designer also needs to relate to the materialisation and the concrete reality of the idea in order to make the solution tangible and manufacturable.

4.1.2 The representational and the non-representational

At the Material Design Lab and in my research, I have witnessed how most design students seem to be 'pre-programmed' with an understanding that the materialisation of a design is a secondary mode in the design process which frequently is only considered at the very end of it. Not surprisingly, when a design is developed without considering or understanding materials or manufacturing processes, it can be very difficult to enter this material mode. This aligns with the findings of the ceramicist and researcher Camilla Groth, who describes her observations of the design process as divided into *representational* and *non-representational* modes. In the *representational* mode, the activities are linked to the immaterial conceptual or imaginary, whereas in the *non-representational* mode the activities are linked to the bodily manipulation of materials and the concrete reality. She states that the designer must move in loops from one mode to the other, but also that moving from representational to non-representational can be abrupt and problematic, especially if the designer has stayed in a representational immaterial mode for too long.

Groth also makes the very valid point that when the designer has had previous bodily experiences of making and manipulating materials, the imagining of a design includes a mental crafting that relies upon these experiences. She therefore concludes that when a designer has this material understanding there is embodied knowledge involved, even during the representational mode (Groth, 2017). Still, this understanding of separation between modes in the design process is fundamentally based upon the idea of dualism, with one side representing the creative mind, the intellect and the imagination, and the other the hands, making and reality. Interestingly, when a design process includes a material dialogue from the beginning, not only will the imaginary activities be based on embodied knowledge of materials, but, as I will justify below, there is also creative imagination involved with the concrete material reality. Consequently, in a design process that includes a material dialogue, this separation between modes can potentially be diminished to a point where it is misleading to categorise the activities in terms of this dichotomy.

Design processes are never identical, nor is there a single answer to the question of what a design process that includes a material dialogue looks like or will contain. However, in the following sections the fundamental activities of this type of process are described in detail and in some cases are exemplified with concrete examples from practising designers or from research. Note that in practice the activities need not occur sequentially. Some are likely to be more dominant at a certain point during the process, and some activities may reappear several times. Nevertheless, like a more conceptual and immaterial design process, one that includes a material dialogue is concerned with framing, imagining, developing and representing a new design.

4.2 Framing

Donald Schön describes how the designer in the process of framing a problematic design situation must set the boundaries, select particular things and relations for attention and impose coherence on the situation, which can guide subsequent moves. He describes how framing a design problem is rarely done in one burst at the beginning of a design process (Schön, 1988). Design problems tend to be ill-defined and Nigel Cross gives this as a possible reason why designers end up working with a 'co-evolution' of both the problem and the solution. He describes how that designers have a tendency to jump to ideas for solutions or partial solutions, before having fully formulated the problem, but also states that in various studies 'expert' designers are repeatedly found to be pro-active in problem framing by actively imposing their view of the problem (Cross, 2006). Thus, a design process will almost inevitably require both analysis and synthesis and during the process of the framing of a design problem, this is particularly evident.

A design brief is likely to influence the framing at the beginning of a design process. A brief can be very simple, or it can include a large number of constraints. Although elements in different design briefs may repeat, and the designer over time gains expertise as a professional, each brief outlines specific constraints introduced by stakeholders, systems or the designer.

It is essential to understand the design problem – including within a design process that comprises a material dialogue; it is the framing that will give the first indications of which type of materials to source. Even if no specific design has been thought of, the framing will define restrictions and requirements for the material. For example, if the brief requires a design for the inside of an incubator for premature babies, the requirements for the materials are perhaps that they must be soft to the skin, breathable, easy to clean, non-toxic, durable and insulating. On the other hand, if the brief requires a design for takeaway food packaging, the criteria may be very different. The designer must study and explore different materials to find the most suitable. The source of the material must be investigated; in addition, the connected traditions, history and cultural heritage, and, equally importantly, the composition of the material must be defined.

4.2.1 Material studies and exploration: the source

In principle, any material could be used for a material dialogue. Still, it is important to remember that although a material dialogue potentially can provide the designer with information about the material that is required to design for sustainability, it is still up to the designer to make deliberate choices (this will be discussed in more detail in Chapter 6). The first time this is likely to be necessary is when the primary materials are sourced. During the initial studies of a material, many things must be exposed: the origin of the material, the abundance, how it is excavated, grown or otherwise produced – and by whom. This information is essential in deciding whether the material is adequate, from an ethical, social and environmental point of view.

For more practical reasons, it is also necessary to study the supply chain, especially if the availability of the material is seasonal or restricted in other ways. Investigating the current usage of a material will help assess its future potential. This often means looking at industries that are not necessarily related to design, such as the food industry, agriculture or the medical industry. In the case of new materials or new material technology, relevant information still might only be at a research level and thus it can be necessary to search for information outside of the field of design.

To maintain material circularity, it is essential to address the potential future of a material, not just in the present design, but also as a material in future designs. Currently, many materials that are in circulation are so polluted or mixed in such a way that, at least with today's technology, they cannot be separated. Consequently, these materials are unable to circulate within a biological or a technical cycle, and thus are not well suited for sustainable design. In order to fully assess the value and compatibility of a material, more aspects must be explored. As will be addressed in the following, some of these are related to experiential qualities and assessing others will require a basic understanding of different aspects of natural science.

4.2.2 Material studies and exploration: tradition, history and culture

Research into the history of a material, and the traditions relating to its use in different cultures, is a way to learn about the meaning of a material. We attribute value to materials: some we associate with high quality, strength or luxury, and others we think of as cheap or even repulsive. These meanings are not necessarily rational, related to the monetary value of the material or even its mechanical properties. They are subjective ideas of materials that are so engrained in tradition and culture that we rarely question them. Furthermore, as they are culturally determined, the meanings of materials are not universally agreed upon and cannot be deduced from theory (Miodownik, 2003).

As designers, we may think of ourselves as neutral, but rather we are just as much a part of tradition and a culture as any user we may design for. We are not necessarily aware of how this affects our decisions and opinions and I have frequently met students who, at least initially, find it repulsive to use glue made from animal bone as a natural biodegradable binder but at the same time they wear leather shoes and defend leather as a high-quality material. Challenged to give reasons for not using glue made from the bone of an animal, when we use their skin, they have no rational explanations, but only answers based upon emotions determined by culture and traditions. Nevertheless, when it comes to the acceptance of a material and a design, emotions are often more important than rationality (Csikszentmihalyi, 1991; Harper, 2017; Karana, Hekkert & Kandachar, 2010). This forces the designer to be particularly aware of the meaning of materials when designing and it calls for an ability to recognise and challenge an initial feeling of rejection towards an otherwise suitable material. This is especially important because a design process that includes a material dialogue is potentially open to all sorts of materials and not necessarily just those that are standard, prefabricated or already familiar to the users.

Some of these sources can be materials that have no meaning for many users, or, even worse, that users associate with poor quality or even find repellent. This is a significant challenge to designers, not least in the transition to sustainable design, where more products may have to be made from biomass waste and recycled materials. Clearly, the designer must explore materials with suitable mechanical properties, but also carefully consider the meaning of a material in relation to the intended function and the user. To change a user's perception of a material requires a designer's profound sensibility for manipulating the material aesthetically.

Another type of information that can be extracted from historical research is technical knowledge. Cultures where a type of material has been dominant for centuries, and where it is still being used, can hold a wealth of relevant information. The recent book *Lo-tek: design by radical indigenism* by Julia Watson shows examples of how traditional techniques and ways of manipulating a material can be relevant for modern sustainable fabrication (Watson, 2019). Sometimes going back in time can lead to answers for the future, but the embodied knowledge involved in traditional craftsmanship and making was typically passed down from one generation to another by apprenticeship and practical training and old techniques were therefore rarely scientifically explained or even written down. Thus, when a certain type of craft knowledge or skills were no longer used, it would only take a couple of generations for them to be lost. Occasionally some knowledge has been recorded, though.

An example of this is the book from 1905 on how to make binders by Ferdinand Dawidowsky (Dawidowsky, 1905). Today, the adhesives used in the manufacturing of products from natural materials are almost exclusively synthetic, petroleum-derived thermosetting adhesives. There are good reasons for this: the synthetic adhesives have been developed for almost any situation and are easy to use. However, this way of mixing natural materials with synthetic non-biodegradable binders can represent a barrier to material circularity and can cause damage to the environment. Prior to the 1920s these adhesives had not yet been invented, and the binders and finishes used in crafts such as cabinetmaking were almost entirely natural based. It is still possible to buy certain types of natural glues, like bone glue or rabbit glue based on proteins from animals, because these continue to be used in the restoration of antiques and fine art. The designer can utilise them in their present form, which has barely changed for the last 100 years and therefore the application process can be rather laborious in comparison to modern synthetic adhesives.

As this example illustrates, it is sometimes necessary for the designer to go back to the past in order to find answers for the future. The other option is for the designer to study *why* the bone glue binds, study recent research on natural protein-based binders (Hemmilä, Adamopoulos, Karlsson & Kumar, 2017) or reach out to material scientists or start-ups for information or collaboration. Nevertheless, the latter option leads to what may be one of the more difficult aspects of material exploration for designers, namely understanding the composition of the material.

4.2.3 Material studies and exploration: the composition

Material scientist Mark Miodownik has repeatedly addressed the importance of promoting an awareness of the cultural impact of materials to the science community, but he also suggests that in order to collaborate it is likewise crucial to promote the science of materials to the arts community (Miodownik, 2003, 2005). In order for a designer to fully understand the potential of a material and its compatibility with others, without merely exploring it exclusively on the basis of trial and error, it is necessary to study the material in greater detail than can be observed with the naked eye at macro-scale. Therefore, the designer needs to have a basic understanding of the science behind the material: to be able to relate to the material at micro-scale, and to have a fundamental understanding of the chemical elements that constitute the material, its composition and structure. The designer should not aspire to become a material scientist, but must have sufficient knowledge to be able to talk to, and work with, a broad range of specialists in the natural sciences.

This is not necessarily a part of existing design curricula, and at present many designers may not have any prerequisites for understanding the composition of a material at this level other than what was taught to them before starting their design education. As I will address in greater detail in Chapter 5, understanding the composition of a material and learning to work with materials can represent a challenge for the implementation of a material dialogue in design education. This is not just because it may be outside the knowledge and capability of the faculty, but also because working with materials in this way will potentially require changes to the physical workspace.

Even when design students have access to a large material library, a decent collection of books on material science and chemistry, as well as the materials database CES EduPack (Ashby, 2008), as they do at the Material Design Lab at Copenhagen School of Design and Technology, this may still not be enough to teach them *how* to study the composition of a material. Because just as with exploring materials at macro-scale, embodied knowledge and practical skills are involved in exploring materials at micro-scale, and these are difficult to learn solely from books and theory. Using microscopes, incubators and pipettes requires practice and often even the types of fine motor skills utilised when handling different tools for craft. Although an experienced chemist or biologist adheres to strictly natural scientific procedures, they will often still be able to determine the status of a sample by smell alone, or without measuring be able to feel when the temperature of a solution is right for a certain process. In other words, exploring materials at micro-scale involves contributory expertise and embodied knowledge just like working with materials at macro-scale.

The process of studying and exploring can vary significantly from one material to another: there is plenty of information and data on standard materials like polypropylene or pine wood that are used frequently in the manufacturing of different products. On the other hand, if the material is apple pulp, as can be seen in Figure 4.1, a biomass waste material from the production of apple cider and juice, the situation is very different, and the designer must be capable of studying

Figure 4.1 **Apple leather**
Hannah Michaud was introduced to apple pulp, a waste fraction from cider and apple juice production, when she followed a course at the Material Design Lab. Her investigation and experimentation with the material led her to develop a leather-like material and a series of different products and prototypes made in the material. She has continued the development in collaboration with the Danish Technological Institute (www.dti.dk) and has founded the company the Apple girl (www.theapplegirl.org).
Courtesy: Copenhagen School of Design and Technology, KEA (Top); photograph Hannah Michaud (Bottom).

and exploring the material on her own. Nevertheless, this is not necessarily a disadvantage, because the information derived from the study and exploration of materials can be closely connected to the imagining of a new design, not just by framing a reality for the imagination, but also because it may in different ways inspire the creation of new ideas.

4.3 Imagining

Imagining what does not yet exist involves forming a mental image of something not present to the senses or never before wholly perceived in reality. Although not present to the senses physically, this mental image will still be based upon the designer's embodied knowledge (Groth, 2017). The psychologist Mihaly Csikszentmihalyi states that a person who wants to make a creative contribution must learn the rules and content of the domain by internalising the fundamental knowledge of the field (Csikszentmihalyi, 1996). In a study of expertise in design, Cross asks a designer that he has identified as an expert to 'think aloud' during the process of designing a fastening device for mounting a backpack on a bicycle. From his comments it is clear that his personal experience of biking with a backpack leads him to identify issues that only someone who has had such experience might be aware of (Cross, 2006). As a result, the designer's imagination of ideas is framed by an embodied understanding of what it means to cycle with luggage.

The activity of imagining what does not yet exist is always inevitably framed by the designer's own experience and understanding of a field – simply because our understanding of the world is not an abstract proposition but fundamentally depends upon our multi-sensory experiences within it (Ackerman, Nocera & Bargh, 2010). Therefore, in a design process that includes a material dialogue, the information derived from studying and manipulating materials will form part of the framework for imagining a new design. Tim Ingold states that it is our ability to give form to something in flux, framed by our experiences, which makes us creative (Ingold, 2013).

4.3.1 Creativity

Creativity has been linked to the mind as an intellectual activity, but research from several fields has by now shown that this is an inadequate explanation (Tanggaard, 2014). In a study of apprentices doing vocational training the psychologist Lene Tanggaard Pedersen noticed how creative they were when allowed to experiment relatively freely with the materials within their domain and make their own projects and designs (Pedersen, 2008). She describes how creativity is sparked by a curious and open-minded enquiry in situations requiring us to respond in new ways (Tanggaard, 2014). Furthermore, she talks about the resistance from the material and refers to Tim Ingold and Elizabeth Hallam, who argue that creativity is a relational phenomenon building upon what humans do, but also what tools, materials and artefacts invite us to do (Ingold & Hallam, 2007). Ingold later described how it is the engagement with the work and not the form of the product that the designer has in mind that creates the work (Ingold, 2013).

Therefore, seeing how a material may compress into a form, how a composite material becomes more flexible by adding a different binder, feeling the softness of a surface, observing the structure of a surface under a microscope or testing techniques that were used to treat a material a century ago directly inspire new ideas. It is the very interaction that a designer enters into with materials and tools that sparks the creativity (Ingold & Hallam, 2007). The quality of the information retrieved from a material dialogue will depend on both making skills and the rigorousness of the material experimentation. Whether the level of creativity also depends on these is more difficult to determine, but Tanggaard Pedersen writes that the resistance from the material being worked with is central for provoking and broadening one's own ways of working. She points out that the resistance is only felt if one dares to engage in immersion with the material, and if a kind of enquiry that is based upon experimenting is allowed (Tanggaard, 2014).

This is the more pragmatic explanation of the creativity involved with a material dialogue. However, as I already described in Chapter 3 about the cognition in craft, in the act of manipulating tools and materials there is a particular bond between body and mind, which can release a mental state of concentration, recreation and even joy. This aligns with the theories of a connection between a feeling of flow and creativity, as outlined by Csikszentmihalyi. He describes that to obtain a state of flow there must be immediate feedback on our actions and we must feel that our skills match the challenge. During a state of flow, action and awareness are merged and we are intensely concentrated on the present and what we are doing. In this state we become too involved in the activity to be concerned with failure and our self-consciousness disappears. With this, the activity becomes autotelic and opens up for creativity (Csikszentmihalyi, 1996).

In a material dialogue the designer is likely to have periods of engaging in craft tasks like weaving, spinning or sanding. In this kind of situation there is an immediate feedback from the material, but the creative state of flow that Csikszentmihalyi describes may also arise in this activity: when the designer is very concentrated on manipulating a material and her skills are appropriate for the given challenge, she may reach a level of automation in the activity that opens up for a feeling of flow and creativity. People who have experiences with knitting, carving or throwing clay for any length of time may recognise how this state of mind can be very compelling. Intriguingly, as I will describe in an example with design students in Chapter 5, this way of sparking creativity does not seem to depend on the level of skills or experimental enquiry, but rather seems connected to the bodily and motoric interaction with the material.

It is not within the scope of this book, nor my capability, to explain all facets of creative imagination that might take place within a material dialogue. What is important is to confirm that there *is* creative imagination involved when a designer is handling materials and tools. The dialogue with the material will both inform and inspire the designer.

4.3.2 Drawing

In the process of imagining what does not yet exist drawing is and has always been central (Cross, 1999; Suwa & Tversky, 1997; Ullman, Wood & Craig, 1990). This was also the case in the beginning of design education during the 19th century. Still, there were at the time different ideas as to how drawing should be used. Apart from being a critic of art and architecture, John Ruskin was an excellent draftsman and he wanted his students to 'see' and draw honest representations of the world. Marion Richardson, a very influential art educator, was opposed to this. She was concerned with eliciting the expression of mental images from her students (Romans, 2005; Ruskin, 1857).

Schön agrees with Richardson by seeing drawing as a design tool for imagining and externalising ideas (Schön, 1983). He highlights how designers can have conversations with their drawings and describes drawing as a reflective conversation with the situation. Like Lawson, he also sees drawing as a tool to inform further design work (Lawson, 2012; Schön, 1983). Drawing can be a tool to document and understand what exists but looking into a designer's sketchbook it is clear how drawing can indeed be used to visualise and advance ideas.

Nevertheless, whereas drawing is a great dialectic tool for imagining that which does not yet exist and can be an effective way of memorising new ideas or solutions, it does not inform the designer about the physical qualities of a design. Drawing is independent of the concrete material reality and as a tool in the design process this is both its strength and its weakness. In a drawing we can illustrate fantasies and the unconceivable, which can spark radical new ideas and different ways of approaching a design solution. However, in design development we can also get 'stuck' in drawing, where our imagination is not restricted by the limits of the material and end up with a design that cannot be manufactured. This is particularly important for students, who are unlikely to have a solid base of embodied knowledge of different materials and manufacturing techniques from previous experiences. This dilemma corresponds with Groth's observations of design students who struggle with bringing their design ideas into a 'concrete material reality' when in the design process they have stayed for too long in a 'conceptual immaterial' mode (Groth, 2017).

At one point, I was so frustrated with my students' tendency to get lost in drawing and concepts that I asked them to refrain from drawing and instead work with 3D sketching directly in the material (Bak-Andersen, 2019). Margaret Wilson, who writes about embodied cognition, argues that we can optimise cognitive work by doing actual, physical manipulation, rather than computing a solution within our heads, but she also points to drawing as a way of offloading cognitive work (Wilson, 2002). My students were not particularly skilled in drawing, but the results of limiting the use of drawing were negative. They appeared constrained without drawing, both in their development of ideas and the design, but also when they wanted to quickly explain an idea or a solution. It was clear that they needed drawing as a tool during the design process for imagining, for memorising ideas, for problem solving, etc.

It is possible that a craftsperson, who has a level of expertise in manipulation of a material that allows for free improvisation, may be able to do most of his sketching in the material. However, this would not appear to be the case for inexperienced designers. Consequently, including a material dialogue does not by default make drawing obsolete as a tool within the design process, but drawing cannot provide the embodied knowledge of the physical realm, which can be derived from material experimentation and the making of prototypes. Therefore, both are needed in tandem throughout in the design process.

4.4 Material experimentation and prototyping

A design process that includes a material dialogue is driven forward by material experiments and prototyping. Although this makes the material central throughout the design process – as it is in a craft process – the designer will not gain the same level of skills and expertise as a craftsperson who has been trained in the same material for years. Yet, through a material dialogue the designer can gain sufficient knowledge of how the material behaves under different circumstances, in order to be able to imagine and develop a new design, either alone or in collaboration with specialists and practitioners from other fields.

Material experimentation and *prototyping* are not necessarily two separate activities; they resemble one another in many ways, and they may happen simultaneously or even in combination. As can be seen in the Figures 4.2–4.4, which show students' experiments and prototypes from a design process that includes a material dialogue, it can sometimes be hard to distinguish between the two. However, for clarity the two activities will be addressed and described separately.

4.4.1 Material experimentation

The material experimentation is primarily concerned with the development of the material. This is not arbitrary but should be conducted within the framework of the design problem. As the process advances and design solutions are being tested and developed by means of prototyping, the material experimentation is likely to become more detailed and specific. At the end of the process, it might even be part of a prototype. Still, the material experimentation is different from prototyping in that it is essentially concerned with the properties of the material. It could, for example, aim at exploring various ways of removing an unpleasant

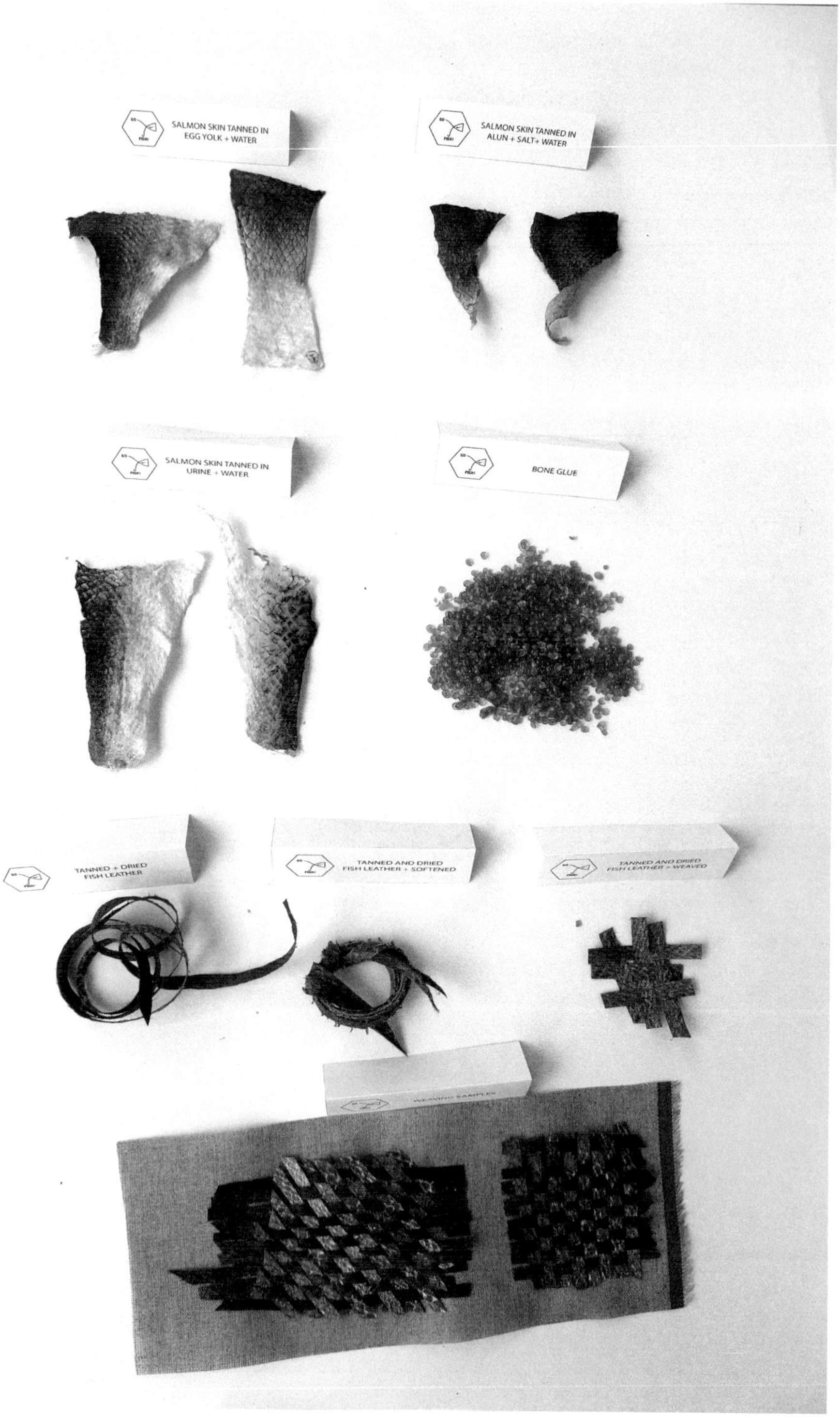
SALMON SKIN TANNED IN
EGG YOLK + WATER
SALMON SKIN TANNED IN
ALUN + SALT+ WATER
SALMON SKIN TANNED IN
URINE + WATER
BONE GLUE
TANNED + DRIED
FISH LEATHER
TANNED AND DRIED
FISH LEATHER + SOFTENED
TANNED AND DRIED
FISH LEATHER + WEAVED

(a)

(b)

Figures 4.2a and 4.2b **Visor cap made from salmon skin (a waste product from the fishing industry). Material samples and prototype**

This visor cap was made from salmon skin, a waste material from the fishing industry. It was tanned using urea (urine), a process the students discovered by studying the history of tanning leathers and fish skin, and it was dyed with natural dyes. The process of achieving a material with suitable properties required a considerable number of experiments. The prototype exhibited on the model demonstrates that the student has good making skills. The material and the function of the product are well matched and indicate that the student has understood the qualities of the material. In general, the prototype represents a convincing design artefact.

This project was part of a four-week collaboration between Nike and the Material Design Lab and the students participating were from the sixth semester in Sustainable Fashion at Copenhagen School Design and Technology.

Courtesy: Copenhagen School of Design and Technology, KEA.

(a)

(b)

Figures 4.3a and 4.3b **Shoe made from recycled wood waste. Material samples and prototype.**
This project explored waste wood and discovered that wood could be turned into many different materials. The participant combined the wood fibres with natural rubber and created a prototype of a shoe. The material exploration and development were both thorough and inventive and several of the material experiments and early prototypes represented valuable epistemic artefacts. Walking in the shoe (the prototype held by the model) was demonstrated. However, the function of the product and the material, in the presented state of development, were not ideally matched, or at least represented a gap of material development before the design could be taken into production.

The project was part of a four-week collaboration between Nike and the Material Design Lab and the students participating were from the sixth semester in Sustainable Fashion at Copenhagen School Design and Technology.
Courtesy: Copenhagen School of Design and Technology, KEA.

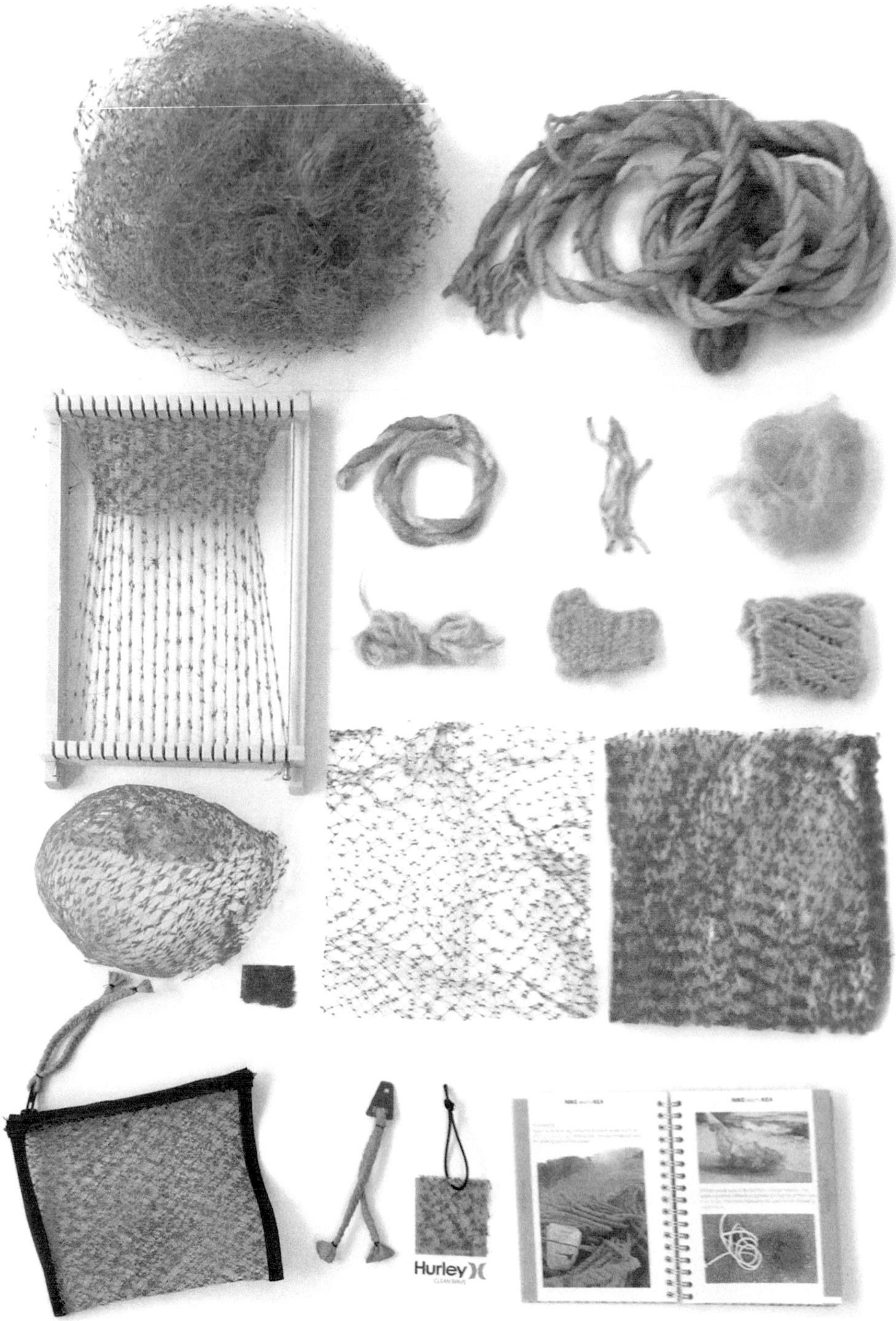

(a)

(b)

Figures 4.4a and 4.4b **Bathing suit made from recycled ocean waste. Material samples and prototype.**
The student behind this project studied and experimented with nylon waste from the sea and subsequently made a bathing suit from recycled nylon. The student had neither the time nor the machinery to produce a high-quality yarn for weaving from the nylon waste; companies, such as Patagonia, were at the time not yet recycling nylon from the sea into new products; and recycled nylon fabric was not commercially available. Therefore, the student made the prototype worn by the model in regular nylon. The material would most likely behave and feel like the recycled nylon and the design was based upon her knowledge from the study and exploration of nylon waste recovered from the sea. However, for the prototype to work as a proof of concept and a convincing artefact, it would in this case have to be accompanied by the material exploration. The product is further away from production than if a standard commercially available material had been chosen, but it brought attention to an (at the time) untapped resource.

The project was part of a four-week collaboration between Nike and the Material Design Lab and the students participating were from the sixth semester in Sustainable Fashion at Copenhagen School Design and Technology.
Courtesy: Copenhagen School of Design and Technology, KEA.

smell, adapting the material to suit a specific tool or applying different binders in a series of material samples in order to test mechanical properties in the tensile testing machine.

Of course, the work involved with the experimentation can vary significantly in both quantity and content depending on the material. A designer who works with

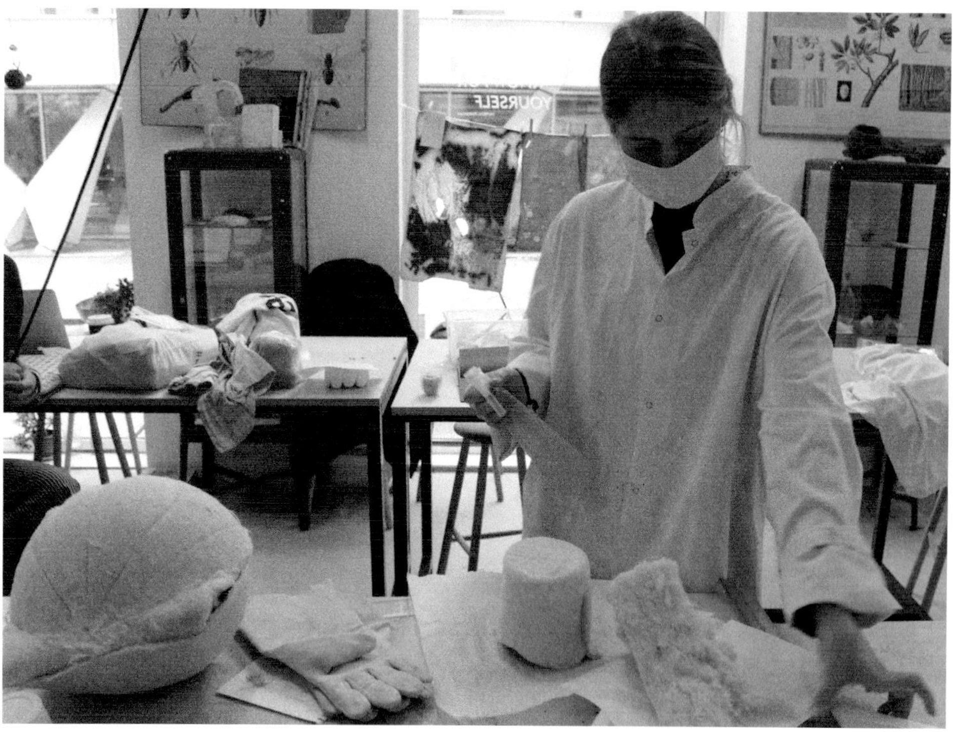

Figure 4.5 **Experimenting with feathers, a waste product from the poultry industry**
This student is experimenting with feathers provided by the company Aeropowder. In the prototype to the left, the feathers have been combined with biodegradable plastic powder and heat pressed into a shape, where the material's potential and insulating properties as a tea cosy are being tested.
Courtesy: Copenhagen School of Design and Technology, KEA.

a more unexplored raw material like chicken feathers, as in Figure 4.5, will likely spend more time on the material experimentation than a designer working with a standard material that is already commercially available, unless this is used in a completely new way. A design process that includes a material dialogue can be described as a series of material experiments where the results and reflections from one experiment inform the next, and in this manner it drives the progression of the design process. This is not unlike the method that has been termed 'serial design experimentation' in design research by Peter Gall Krogh, Thomas Markussen and Anne Louise Bang. They describe the method as 'experiments carried out in a certain order or logic of locality determined by how neighbouring experiments in a sequence influence one another' (Krogh, Markussen & Bang, 2015).

The individual material experiments can be defined according to their purpose, which generally addresses either *technical* or *experiential* features of the material. *Technical* experiments are concerned with measurable qualities such as strength, durability, flexibility, waterproofing, toxicity, biodegradability, recyclability, weight, conductivity, etc. *Experiential* experiments are concerned with subjective qualities such as aesthetics, smell, sound, feel and cultural meaning and perception. To work this way requires a well-developed understanding of experimental practice: the ability to set up, conduct and document different types of experiments, know how to extract knowledge and new questions in order to proceed. As I will describe in Chapter 5, this does not come automatically, but has to be learned.

4.4.2 Iterative prototyping

Prototyping is the act of making a preliminary version of a product from which later designs are developed. A prototype is an artefact that designers can use to evaluate potential design concepts before proceeding with further development (Häggman, Honda & Yang, 2013). Although many people may see prototyping primarily as a way to communicate a final design, rather than as a tool in the design process (Kolodner & Wills, 1996; Lawson, 2006), mechanical engineer and professor at Massachusetts Institute of Technology Maria C. Yang argues that the most important outcome of prototyping is the knowledge that the designer acquires during fabrication (Yang, 2005). Iterative prototyping can be a way to develop a product within a cycle shaped by designing, building, testing and learning.

During a study of prototyping in engineering schools in Mexico, Schlecht and Yang observed how one of the involved groups of students appeared to have an easier time adapting to the constraints because the materials were incorporated into their "design world" from the beginning, before they had time to become fixated on a particular design or process and, thus, they were more open to unexpected ideas. They suggested that this group of students may have performed better because they accepted the provided materials as part of their design space, rather than over-constraining themselves too soon. The participating students in the study commented how the material constraints seemed to help "ground" them by providing specific materials in their ideation (Schlecht & Yang, 2014).

Depending on the type of prototype being made, it will provide the designer with different kinds of information (Houde & Hill, 1997; Ullman, 1992). There are various ways of classifying prototypes. They can be thought of in terms of their purpose or the categories of questions that they answer about a design, or one can classify them according to the stage of development (Yang, 2005). These questions may be more or less defined and *can* be set up to provide quantifiable data. However, they are more likely to be of an explorative character typically beginning with a 'what', 'why' or 'how'. Consequently, within the context of a material dialogue, it makes sense to describe prototypes in relation to the purpose that they fulfil.

What can be tested and developed by prototyping in a material dialogue can, like material experimentation, be divided into *technical* and *experiential* features. The *technical* relates to aspects such as function, assembly and disassembly, structure, usability and form, in relation to ergonomics. The *experiential* qualities that can be explored in a prototype are aspects such as the user's perception of the product, the style, the feel, the colours and the form in relation to aesthetics. Even a prototype built to test one thing may answer other questions than those originally asked (Yang, 2005); thus, the information extracted from a prototype is likely to be of both technical and experiential character.

In an earlier study, Yang describes how that the building of prototypes can be seen as a trade-off between fidelity of the prototype and the resources required to produce the prototype, including time, effort and cost. This means that designers should choose the simplest prototype that can be built quickly and inexpensively, but that still provides the desired information (Yang, 2004). Within a contemporary context, prototyping does not necessarily include the materials intended for the final design but may be made of any representational material. Using the real materials may in some cases require more work, but it will provide the designer with feedback and inspiration framed by the concrete material reality. Using a model building material like foam to make a prototype may inform the designer about lines and shape, but it will not test if the real material is suitable for this shape.

Prototypes can be built at a reduced scale. This kind of prototype may be useful for testing principles for assembly or spatial proportions. However, if a prototype is built at a reduced scale in the real material, it may misinform the designer on both experiential and technical aspects of the suitability of the material. A full-scale prototype that is built in the real material will provide the highest fidelity. However, some enquiries can be tested with less, and even in a design process that includes a material dialogue, it is not *always* possible to make a specific prototype in the real material.

Several researchers have documented the value of prototyping during the design process, and studies of practice suggest that for a successful outcome, designers should rapidly make multiple prototypes early on during the process in order to quickly test design concepts and make modifications (Schlecht & Yang, 2014). Furthermore, simulating a design by means of prototyping has been seen to reduce design risk without committing to the time and cost involved in full production (Houde & Hill, 1997). This follows former professor at the Bauhaus

László Moholy-Nagy's ideas about prototyping; already as far back as 1923, he stressed that an economic production is only possible with the aid of the prototype (Moholy-Nagy, 1923). Perhaps this is because a well-developed prototype is a proof of concept, and maybe also because the prototype is a convincing tool for communicating a design. Presumably for similar reasons, projects without a working prototype are rarely funded on the incubation platform Kickstarter.com (Camburn & Wood, 2018).

The prototypes made during the design process will have different levels of completion and detail. Early prototypes may be very simple mock-ups, but further into the design process, the prototypes are likely to be more elaborate and go through iterative stages of building and evaluation. At the end of the process, there is likely to be a 'final' prototype that represents the materialised design idea as close to the finished manufactured product as possible. Prototypes at all stages of the design process have, like the samples from the material experimentation, potential as a conversation piece when trying to solve specific challenges. They may be used in co-creation with manufacturers to discuss fabrication, with material scientists to talk about specific challenges, or simply to get a response from a user about their opinion of a design when a material is applied in a novel way.

4.4.3 Epistemic artefacts

The material experiments and prototypes are artefacts in a design process, therefore the most obvious way to evaluate them is as design artefacts, considering aspects such as form, function and not least aesthetics. However, they also have the purpose of generating knowledge and, interestingly, the most successful design artefact is not always the one that will produce the most profound knowledge or push forward the development of the design. To understand this, it is valuable to look at *epistemic artefacts*, known from design research.

Artefacts often play an important role in practice-based design research (Nimkulrat, 2009, 2012). In the dictionary, 'artefact' is defined as 'anything made by human art and workmanship; an artificial product' (OED, 2017). However, as Risto Hilpinen specifies, an artefact is defined by purpose and, thus, things made unintentionally, like sawdust, cannot be categorised as such (Hilpinen, 2011). It is also in the purpose that the design artefact and the epistemic artefact differ. The design artefact is an object conceived and planned by a process of design and it makes sense to evaluate it as such. The sole purpose of the epistemic artefact is to produce knowledge (Hansen, 2010), and therefore it should also be utilised and evaluated differently.

The role of the epistemic artefact is to relate to the developed knowledge. It is a tool to communicate and discuss knowledge produced during the design research; therefore it can be considered successful when it evokes new questions and inspires the making of new experiments (Hansen, 2010). This is why a good design artefact is not necessarily also a good epistemic artefact. The artefact made by Maria Sparre-Petersen in Figure 4.6 is the result of an experiment

Figure 4.6 **Epistemic artefact composed of glass blanks fused together**
The Epistemic Artefact CK8_2016_12_B is part of the project 'Composition # Devitrification', which builds upon the results from the experimental process of the PhD project 'Recycle: about sustainability in glass craft & design' (Sparre-Petersen, 2017). A central agency of the thesis is the establishment of a foundation of circular principles for carrying out craft and design work in glass. The glass blanks were handmade from recycled container glass. The container glass has the capacity to devitrify more than the semi-crystal glass traditionally used in lifestyle products. The devitrification transforms the glass from an amorphous to a crystalline structure, which adds opacity to the surface of the glass and highlights the shapes of the blanks in a way that emphasises and accentuates the pattern composition of the finished artefact.
Courtesy of designer Maria Sparre-Petersen.

conducted primarily for knowledge, but which shows that an epistemic artefact can be aesthetically attractive and informative at the same time. Nevertheless, new enquiries and interesting findings can come from material experiments and prototypes, which as design artefacts can be seen as less successful or even failed, while a visually unappealing epistemic artefact may be still be highly informative.

This is important to remember when we introduce a material dialogue in the design process, because it will likely result in a design process that, at least at times, is less clean and aesthetically pleasing than a design process that is expressed in concepts and digitally created visualisations. Thus, in a design process that includes material experimentation and prototyping, we must remember not to pay attention just to the *successful* design artefacts, because, as Figure 4.7 shows, an *unsuccessful* design artefact may in fact be a very good epistemic artefact. The epistemic artefacts are valuable physical telltales and important tools to communicate knowledge and generate new enquiries. Consequently, in design research epistemic artefacts can be a useful vehicle for theory development, but if we appreciate their value and use them in design practice, they can also push forward the development of a design.

4.5 Conversations with designers

A design process that includes a material dialogue is simply a different way of designing. To some designers it may seem radical and difficult to grasp, to others it will be similar to the way they are working at present. It does not differ in purpose and, as described in this chapter, it is still concerned with framing, imagining and developing a design. However, introducing a material dialogue potentially alters how we approach these different tasks. As with any other design process it will be defined by the individual designer's skills, knowledge and ambitions, and it will adapt to the specifications from different design briefs and therefore vary in focus, methods and collaborations. The central elements that characterise a design process, which includes a material dialogue, are the following:

- The principal materials aimed for production are sourced in the initial phase of the design process.

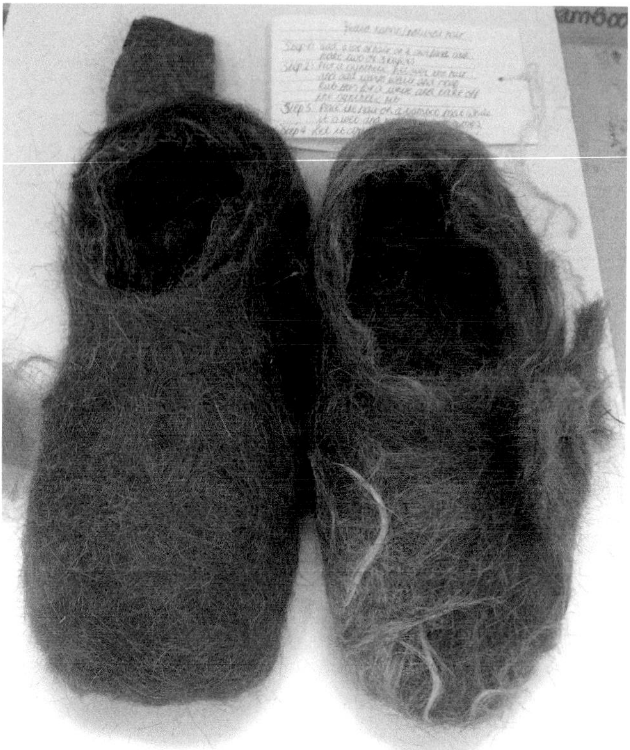

Figure 4.7 **Human hair slipper**

This prototype of a human hair slipper can in many ways be seen as a failed design artefact. The hair sticks and feels itchy, it is made from a material that most people for cultural or historical reasons find unacceptable and the design is in itself not particularly convincing. However, as an epistemic artefact it is very informative. It exposes a challenge for designers to work with rejected materials that are readily available to us, which requires an incredible sensibility, and it is a good example of the fact that not all aspects of sustainability can be measured. This shoe would present a very convincing score in a lifecycle assessment. Nevertheless, it is evident that this is not a product that many people would keep. Partly because it is uncomfortable to wear and partly because many would find it repulsive to wear a product made from human hair. Cultural rejection of a material, comfort, aesthetics and perceived value are all very relevant for sustainability, but they are hard to measure and therefore not exposed in a quantifiable assessment.

Courtesy: Copenhagen School of Design and Technology, KEA.

- The materials are explored and studied, in order to gain a full understanding of the source, the technical aspects, the composition, the historical and cultural heritage, and the meaning of the material.
- All development of form, function and aesthetics is informed, inspired and restricted by the material.
- Material experimentation and iterative prototyping are central design tools in the development of the product.

By including a material dialogue into the design process, the designer can obtain a profound understanding (and thereby potentially also the control) of the physical realm of the end product. The sharp division between the mode of purely conceptual creative thinking and the mode of acting in concrete material reality no longer exists, because the imagination and the development of the design are framed and inspired by this reality.

To exemplify how designing with materials may be expressed in a designer's practice, and to provide more detailed and individual insights, it is worth sharing two interesting conversations with very capable and recognised designers. The conversations were set up as simple semi-structured qualitative interviews that seek to comprehend and explore the subjects' understanding of a topic. The expert practitioners are asked to describe as precisely as possible their experiences, their actions and reasons regarding their way of designing, to obtain descriptions that are as inclusive as possible (Brinkmann & Kvale, 2015). As the experiences are individual, the two conversations have the same starting point and focus, but do not follow the same line of questions. Brinkmann and Kvale describe this epistemological conception of interviewing with the metaphor of the interviewer as a miner, who attempts to collect knowledge and data out of a subject's experiences.

4.5.1 Interview with designer Jonas Edvard

Could you please introduce yourself?

I am a Danish designer, I have a Bachelor's in industrial design and I graduated with a Master's from the Royal Danish Academy of Fine Arts, School of Design in 2013. The focus of my work is investigating the potential of new materials, especially from natural resources, and applying these materials into a functional object. I want to develop new sustainable products that can both explore and explain the possibilities of building our planet in an environmentally friendly way. I am in general very concerned with embracing nature.

How did you learn to design this way?

When I started design school, my initial interest was not sustainability, but the composition of materials. This interest did not fit with the design education at the time. At the Bachelor programme in industrial design no one was working with material-based design. There was a focus on problem solving, but we were taught that a design process should start with a drawing and then you would make a prototype – although, sometimes we did not even do a prototype. I rapidly became tired of starting with a drawing on a piece of paper, but I felt very alone with my interest in materials. I ended up switching between departments, because my approach to design and materials was not acceptable. In some ways this was good because I ended up trying all the traditional craft disciplines – glass, ceramics, wood, metal and textiles – but my approach to design was not really fitting in. I wanted to challenge the material, make experiments and prototypes.

This was a new perspective on design, a different way of designing. I felt I met a barrier at school; I wanted to challenge the way we learned to design. I ended up in the ceramics department, where I met some teachers who understood me and could help me, and in the ceramics workshop I was allowed to experiment with materials. At the time when I was graduating, I was told that I should not stay all the time in the workshop, I had to base my design on academic research. At that time there was a big change in how we were told to approach the design process. I felt this was wrong. Not because it was academic, but because the teachers demanded an academic approach before we were allowed to experiment. I think the best way of developing ideas is by letting go off my thoughts through investigating and 'play around' with materiality, at the same time as doing research. In this way you construct materials and thoughts at the same time. I do think that this is changing and that some schools, including the school where I studied, are starting to implement some degree of material exploration in the earlier phases of the design process and also letting the students use their hands to learn about materials and processes.

During my education I felt very alone working with materials, but later I learned that there were in fact other designers who were working in that way. Now there is a movement where more designers are working with materials and some design schools are also opening up for a different approach to design. Many designers are focused on making and using new materials from natural resources and waste. Of course, the goal is to have a clean society with no waste, but working with waste can be incredibly complex, because you might think that you are solving a problem by making a new product from waste, but unless you are very careful and understand the circularity of the material fully, you may in fact be making the problem bigger. In design we seem to have agreed which materials are suitable for what, but this means that we overuse the traditional materials like ceramics, plastic, glass and wood and when we mix them we pollute them in ways that they can no longer be recycled. So, I was happy to find out that there were other designers who wanted to change the way we design things, but I still have the feeling that my way of designing is different from most designers.

Could you talk about your design process and the role of materials in it?

I think experimentation is the key to understanding what a new material and processes can do and bring into the discussion of sustainable products and how we use those products. Material experimentation opens up a landscape of ideas; it gives inspiration, which eventually can become a product. I typically call this 'motivation'. It is a feeling that makes me want to investigate and explore something. When I experiment with materials, I write everything down – weights, temperature – all relevant parameters. I follow a plan for testing and use a scientific approach to material investigation. I explore new ways of doing things by testing. It can be quite boring to do all these tests, but suddenly when I strike a result it is all worth it. I tend to start seeing patterns in the tests and I will begin to discover new aspects and get unexpected results. It shows me a different side of what I am doing, something unpredictable and surprising, and it almost feels magical. It tells me that here is something I need to investigate further, something new and original. This is the best situation for me as a designer. I think designers can learn a lot from a scientific approach to experimentation. But designers are fortunate because they also have the possibility of using an artistic approach, which is somehow wired differently and is far from the way a scientist would work. But it is very useful, because it is connected to emotion, to reflection and behaviour, which are very important elements in the designing of everyday things.

How do you choose a material?

I fall in love with some materials and processes – I become fascinated with things I don't understand. I don't really read papers about materials: I look at them; I dive into them. I like to have a lot of materials around me. Often, I will find that a product could be improved by taking one material and replacing it with something else. I look at composites and question why they are made the way they are and consider if they could be made differently. We did a project designing sound-absorbing panels for a company and we wanted to find a natural fibre which could replace fibreglass. But when you approach natural materials there are so many different materials and frequently they have not yet been transformed into a standard commercial material, so it may come in the form of a loose fibre, just cut and green, or be spun or woven, or even as a waste material and broken down into bits. The origin is important: the closer the source the better. So even if I would love to work with coconut fibre, I don't, because it is not a local material. I find materials most inspiring when they are accessible, abundant and at hand, something I can find or harvest locally. I don't get inspiration from something that I read about on the Internet and need to order from far away.

What should design students learn?

This ability of a designer to understand and work with materials is important in a design process, because if you go to a shop and buy a standard material it is made for standard manufacturing. But design students must learn that they do not need to accept the material as it is. They can be part of forming the material, ensuring that the material and the product are sustainable. As a designer you should start the design process before there is a drawing, you must look and explore materials, because they all have different characters and value and the material brings something into the design process that is unique. It is the responsibility of the designer to enhance and use these qualities. I think over the last ten years the focus on materials has increased, but I don't think design students really learn enough about experimentation.

When I teach, I sometimes give the students the task of doing material research in a specific area, and I ask them to record all the material sources they find and make a small catalogue where they describe the materials. This description should not necessarily include measurable facts about the material, but more about the origin, the narrative of the material as well as its meaning and emotional value. It is important for designers to understand the value of the materials they use.

Do you think your way of designing carries the heritage of a particular Danish way of designing?

Yes, I experience this every time I go abroad to teach or exhibit and meet foreign designers. I would say that the connection I have to craft is very Danish, my way of looking at the possibilities of a material and testing it. In the history of Danish design there has been a strong connection to natural materials: wood, ceramics and textiles. There is a tradition for investigating colour and material combinations and a noticeable attention to detail, and I feel I am part of that tradition.

How do you see the future of design and design education?

Today designers can make a design on a computer and it looks really original and nice. For several years the argument for designing this way has been that everybody would have a 3D printer in five years, but this has not really happened. I think both universities and designers must challenge the future of design. The question is, what the role of the designer is going to be – just a thinker and somebody who draws ideas, or somebody who actually contributes with sustainable products? It is important to discuss the responsibility of designers, because there is a continuous demand for objects, so designers must ask themselves if their main focus will be to produce a lot of things that fit into the category of 'design objects'.

You talk a lot about material experiments and prototypes in your design process. Could you explain the difference between a material experiment and a prototype?

Material experiments are a series of tests aiming to clarify and expand the knowledge about a certain property of a material. They are a way to test given facts, and a way to explore and define the possibilities of a material or process. We often test materials by mixing and adding chemicals to alter the properties of a selected resource. The prototypes are based on the analysis of the experiments and are a way to test form, functions, details or look. Prototyping involves larger-scale tests and mock-ups, which eventually lead to the final design solution.

More information about the work of Jonas Edvard can be found at www.jonasedvard.dk.

Figure 4.8 **The design studio of Jonas Edvard**

The way a designer works is likely to be reflected in the type of workspace and its set-up. Although Jonas Edvard does have a space upstairs with a large desk where it is possible to sit down, draw and work on a computer, his studio looks more like a craftsman's workshop with elements from a laboratory.

Courtesy: The Mindcraft Project, stills from the movie by Michael Sangkoyo Gramtorp.

(a)

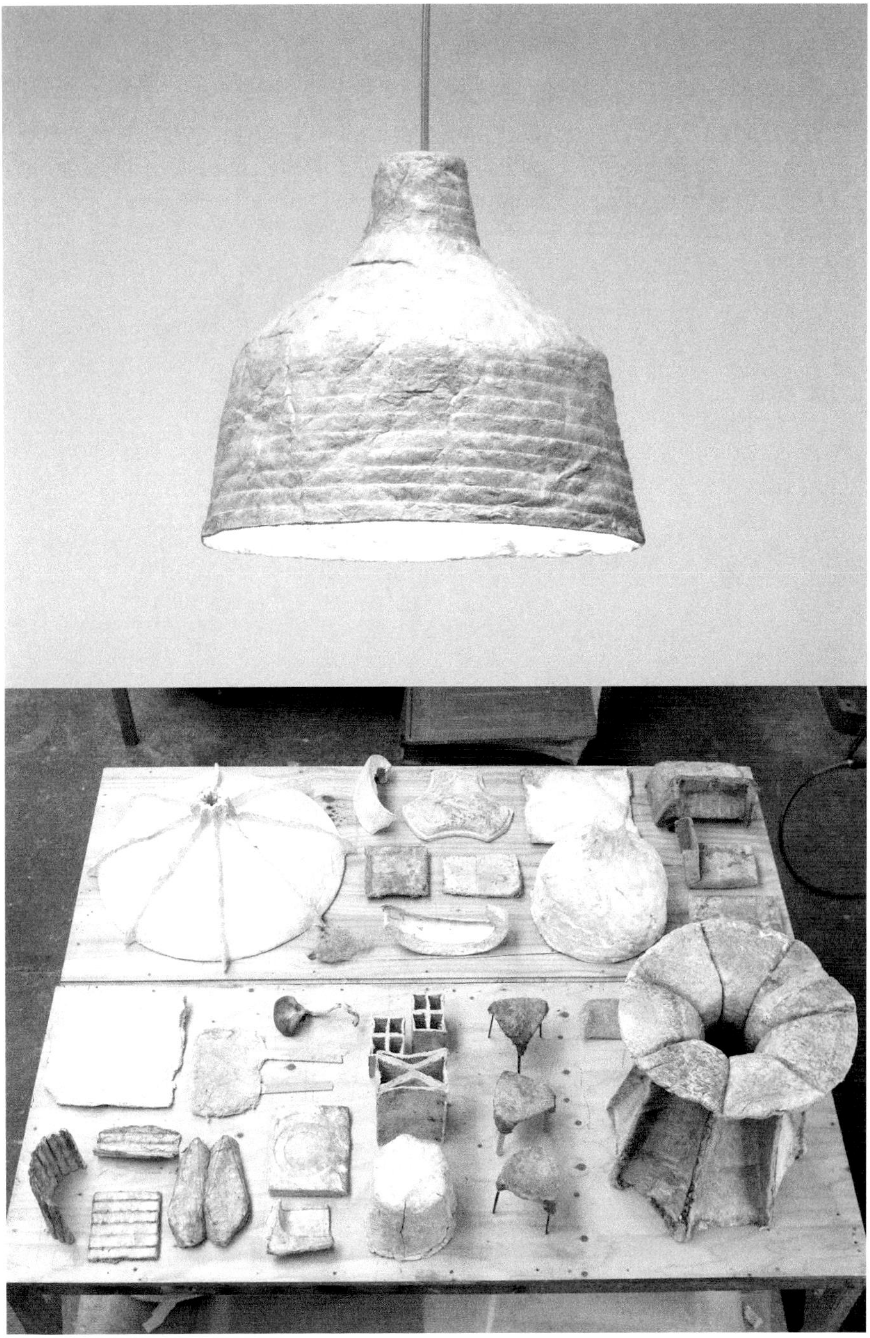

(b)

Figures 4.9a and 4.9b **Material experiments, prototypes and lamp made in mycelium**

Jonas Edvard started working with mycelium in 2013. His approach to material experimentation and prototyping is both artistic and scientific.

Courtesy: The Mindcraft Project, stills from the movie by Michael Sangkoyo Gramtorp (Figure 4.9a); Jonas Edvard (Figure 4.9b).

(a)

(b)

Figures 4.10a and 4.10b **Making the mycelium chair, MYX**

The chair is made from a strong and lightweight mycelium and hemp fibre composite. The mycelium feed on cellulose and grow into the layers of folded hemp fibre matts to make a solid corrugated board. The chair is constructed out of 15 boards that in the end are assembled and grow together. It takes four to six weeks to grow the chair.

Courtesy: The Mindcraft Project, stills from the movie by Michael Sangkoyo Gramtorp (Figure 4.10a); Jonas Edvard (Figure 4.10b); photograph: Anders Sune Berg.

4.5.2 Interview with the designers Maartje Dros and Eric Klarenbeek

The interview is with Maartje Dros, she speaks on behalf of herself and Eric Klarenbeek.

Could you please introduce yourself?

Eric and I are designers, we are based closed to Amsterdam in the Netherlands and we were both educated at the Design Academy in Eindhoven, which is an art-based design academy. Eric was focusing on identity, innovation and robotics and I was focusing on public space and social design. We graduated in 2004 and 2005. After graduating we had an interest in making objects with new types of fabrication methods, so we started to use 3D printers and gradually more machines were brought into our studio. We used them to design and produce locally, but also to get a better understanding of and connection to fabrication processes. The first 3D printers we had in the studio we built ourselves, and that was a good way to learn about production and machinery. Initially we only used plastic filaments, which were available as standard materials for these printers.

What changed, and when did you start designing the way that you do now?

Eric went to India on a field trip and this changed our perspective. He went to a 'recycle city', which is a neighbourhood where everybody is into the recycling of plastic. The locals were living in the workspace for recycling and they stored plastic waste on top of their houses. The activity of recycling was the livelihood of the community, but they were living and working in a very poor and toxic environment. It was an eye-opener, because in some ways we were doing the same: more and more independent designers and makers opened workshops and studios with bad air-ventilation systems filled with Fab Lab machinery, such as 3D printers, laser cutters etc., processing plastic-based materials and filaments. Recycling is in principle good, but we discovered that we need to look at how it is actually done. After this we really started considering which materials we use; we became much more critical of how we use them and the impact they have. We felt we needed to change the way we were working.

We started a collaboration with Wageningen University, which has a strong profile in agriculture. Their interest was in finding ways to push forward innovation in agriculture by integrating additive manufacturing. We were connected to the mushroom research group working with mushrooms and mycelium. They were producing mushrooms on flat surfaces, but they wanted to find ways to grow them vertically. We invented different structures and tested some product ideas, but we rapidly learned that on oil-based non-biodegradable plastics, there was no growth, whereas when we used wood-based materials the mycelium started to grow.

Working with growth in design meant that we needed to understand nature. I think this was the first time we discovered that the information we needed was not available and the only way to get it was to educate ourselves through collecting and combining information from different experts from both art and science. To educate ourselves we had to move everything to our studio: we started growing mushrooms, building and adjusting the 3D printers to suit the mycelium, and we needed to learn how to set up and work in a sterile environment. We learned a lot from people working with aquaponics and home growing. It was also our entry into a network where we could get

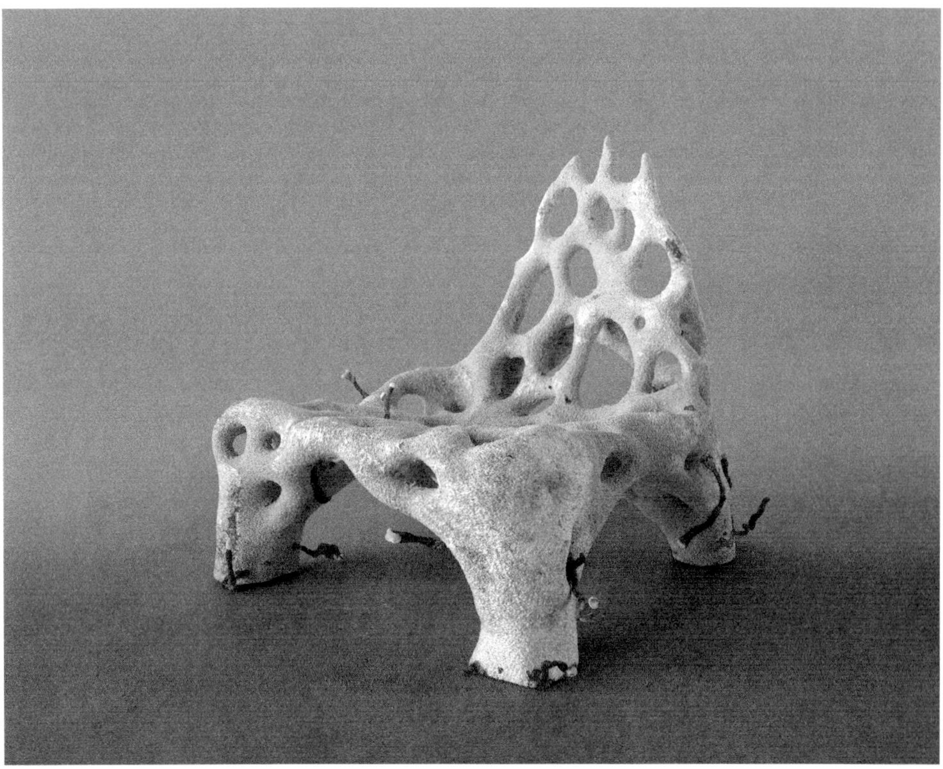

Figure 4.11 **3D printed mycelium chair.**
Courtesy: Klarenbeek and Dros.

cheaper equipment and we would order small parts and assemble them ourselves. The knowledge that we gained in the process opened up the possibility for better communication with our partners. It enabled us to talk about the material with the mushroom farmers and the scientists on a different level.

To start out a project by educating ourselves in order to be able to communicate and understand experts from other fields is a very important part of the way that we work. As designers we have the freedom of not having to follow the same strict scientific rules as they do at the universities. We can work more intuitively, but we can only do this when we know the material really well. This means that we need to explore it, grow it, fail to grow it, test it with different machines and so on. But it also means that we come up with samples and ideas that our partners may never have thought of or considered testing, and in collaboration we can investigate further or push a product forward to a point where it also starts becoming economically and socially interesting. With this way of working we can take responsibility for what we do.

Could you try to exemplify your design process with a specific project?

Eric was the first to 3D print a vertical structure with mycelium, but after printing mycelium, we also became interested in algae, because of its ability to store CO_2 and produce oxygen, and we got in contact with the Dutch seaweed farm. Holland is surrounded by water and there is a big potential for growing things in the sea. An interesting thing about algae is that it can be used to make bioplastics. A research group was looking at the marine environment around the windmill parks in the sea. They wanted to explore and implement sea farming and had started looking into how the seaweed could be used, not only for food, but also for materials. We had the right set-up at our atelier to do experiments and felt it could be interesting to find out if it was possible to make a filament for 3D printing from the algae.

We started a collaboration with the university Avans Breda and they allowed us to work in their laboratories, where we could develop filaments from the algae ourselves. We got the seaweed and the support of an open-minded professor. In Breda we created a first filament with seaweed, saw the potential and started investigating further. Therefore, when French Atelier Luma in Arles wanted to introduce a social design and makers' platform we proposed the Algae Lab, and copied a lab set-up similar to our studio at home and shared our experimental way of working. In Atelier Luma we decided to work with the micro algae, spirulina, which is common and farmed already in southern France. We connected the cultural heritage of Arles with new concepts for labour and produce in the city and made historical utensils available again by printing them on location. This led to a series of 3D printed glassware printed with local algae and inspired by Roman glassware found in the Rhône. Basically, we wanted to make a product that would educate people and show them how it is possible to think and work circularly, by connecting algae from the nearby sea with a product for their homes.

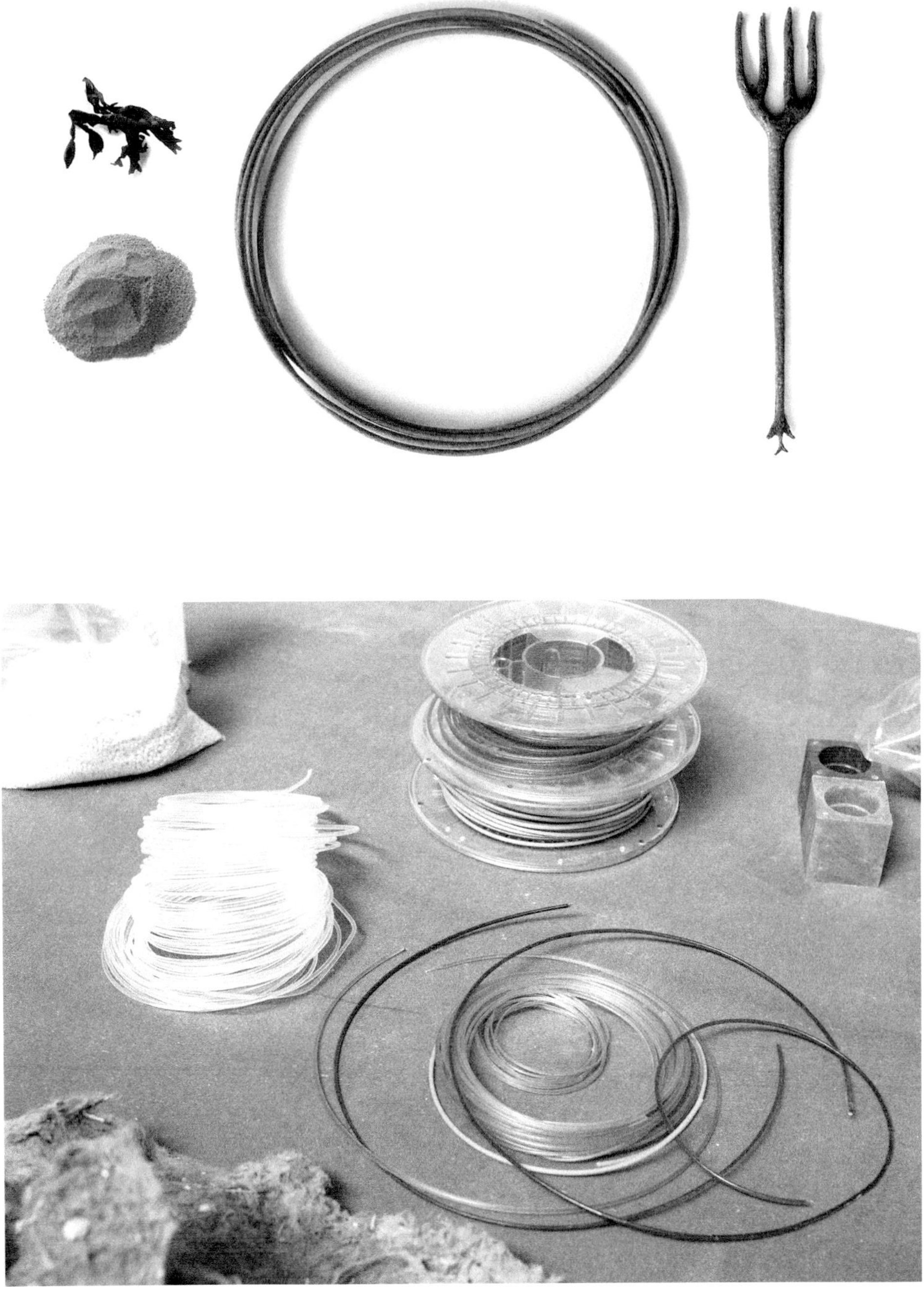

Figure 4.12 **Designing with seaweed**

Material samples and prototype from the process of turning seaweed into a 3D printed object. From seaweed to granulate to filament and, finally, 3D printed fork (Top). Different filaments for 3D printing made from algae (Bottom).

Courtesy: Klarenbeek and Dros; photograph: Victor Picon.

Figure 4.13 **Designing with spirulina.**

Harvesting spirulina samples in southern France (Top). Growing spirulina in a temporary studio at Atelier Luma in Arles, France (Bottom).
Courtesy: Luma; photograph: Johanna Weggelaar (Top); photograph: Antoine Raab (Bottom).

Figure 4.14 **3D printing with algae**

3D printing with bioplastic filament made from algae.

Courtesy: Luma; photograph: Antoine Raab.

Figure 4.15 **The studio of Klarenbeek and Dros**
A view into the studio and workspace of Klarenbeek and Dros, set up to work in individual sterile working environments.
Courtesy: Klarenbeek and Dros.

Does your design process in general start with sketching, material experimentation, prototyping or research?

We basically do it all at the same time. For instance with the algae project, we will grow the algae in our studio, we will dry the algae, grind it down, make it into a granulate, make a filament, we will adapt tools at the same time as we do research, draw, do the 3D printing, go to the museum to scan Roman glassware and talk to the curator, and talk to the farmers, the manufacturers and make prototypes. There is a lot of reflective thinking involved with the process, the material influences the design we propose and vice versa. I think it is interesting to discuss when something is a product. Is it when the material is functioning as algae in the sea absorbing CO_2, when it is on the table being used or is it when it is returned to the ground and functions as a nutrient? I don't necessarily think of our design as material focused, but we work with the materials, because it is a necessity if we want to design for circularity. To design this way, you must get your hands dirty, it has no value only as a concept and you cannot invent this product simply by thinking.

Retrospectively, do you think there is anything that you would have benefitted from learning in design school to be able to design the way you do now?

We were not really trained in documenting our experiments and findings; if we had known how to work in a more scientific way, we might have had a better vocabulary and it would have been easier for us to collaborate and be taken seriously. We see students who want to work with material research and design and they are lacking this. Designing this way is not like making a drawing in your computer and visualising it. That way of working is clean; when you work with materials you get dirty. When I look at workspaces and laboratories in design school, they are not integrated into the education and not available to all students. Working in these spaces seems to be a side-thing and not where the main part of the design process takes place.

I think design education should be more integrated into the students' lives; it becomes a thing of going to class and going home. The situation with Coronavirus makes us aware that there is a lot to learn about design outside school. Going on field trips, learning from real-life situations and working with experts outside school teaches the students to collaborate. When we studied, we did these intense one-week workshops where we were working outside the school and I think these were what taught me the most. You were there, in the environment you were designing for, and you were learning on all sorts of levels in a way that you could never learn from reading a book.

Do you see things changing?

I think there is a growing focus on craft; perhaps we lost the connection to materials and how things were made for a while. In our studio we focus on that, but we also question how things are made. Most design schools have workshops for working with wood and prototypes, some also have Fab Labs, but not many places have spaces that are suitable for exploring materials. It requires different tools and equipment. Working with natural materials you often need a sterile environment. Without suitable spaces it is hard for designers to get a more profound understanding of materials and what they can do. Some universities in the Netherlands are opening up for these kinds of workspaces.

It is also about how you teach – teaching in these kinds of labs requires one-to-one sessions, where you can show the student how to use the space, demonstrate how the microscopes work, discuss when things go wrong and so on. Sometimes when we teach a workshop in the biolab, the students expect that they will learn everything about algae, but this is of course not possible in a short time. I also don't think it is necessary, because what is most important is that they learn to stay playful, learn to do research, to explore the material and use it with different tools. This will enable them to question standard manufacturing processes and find ways to use the material.

Do you think your way of designing carries the heritage of a particular Dutch tradition?

We do have a tradition for collaboration. We found that there can be a strong hierarchy in institutions and companies outside the Netherlands, which can make collaboration difficult. In the Netherlands we have quite a flat structure, which means that it is acceptable for a designer to contact big companies or important professors at the university. We are a small country, so we need to collaborate, and nothing is ever far away geographically. Also, we have always been forced to work with the land and design nature to suit our needs: we have built dikes to protect ourselves and had to develop different approaches to agriculture and construction, so we are used to rethinking the way we do things and we are obliged to work with nature and the environment.

More information about the work of Maartje Dros and Eric Klarenbeek is available at www.dotunusuals.com.

References

Ackerman, J. M., Nocera, C. C., & Bargh, J. A. (2010). Incidental haptic sensations influence social judgments and decisions. *Science (New York, N.Y.)*, *328*(5986), 1712–1715. doi:10.1126/science.1189993.

Adamson, G. (2010). *The craft reader*. London, UK: Bloomsbury.

Adamson, G. (2017). *Thinking through craft* (5th ed.). London, UK: Bloomsbury.

Albers, A. (2018/1968). Interview with Anni Albers, 1968, July 5th. Archives of American art, Smithsonian Institution. Retrieved from https://www.aaa.si.edu/collections/interviews/oral-history-interview-anni-albers-12134#transcript.

Ashby, M. (2008). The CES EduPack database of natural and man-made materials. Retrieved from https://www.grantadesign.com/download/pdf/biomaterials.pdf.

Bak-Andersen, M. (2019). From matter to form: reintroducing the material dialogue from craft into a contemporary design process. PhD thesis, the Royal Danish Academy of Fine Arts, Schools of Architecture, Design and Conservation.

Bing, S. (2010/1895). Artistic America. In G. Adamson (Ed.), *The craft reader* (pp. 99–106). London, UK: Bloomsbury Academic.

Brinkmann, S., & Kvale, S. (2015). *Interviews: learning the craft of qualitative research interviewing*. Los Angeles, CA: SAGE.

Buchanan, R. (1995). Rhetoric, humanism, and design. In R. Buchanan, & V. Margolin (Eds.), *Discovering design: explorations in design studies* (pp. 23–58). Chicago, IL: University of Chicago Press.

Camburn, B., & Wood, K. (2018). Principles of maker and DIY fabrication: enabling design prototypes at low cost. *Design Studies*, *58*, 63–88.

Cross, N. (1999). Natural intelligence in design. *Design Studies*, *20*(1), 25–39.

Cross, N. (2006). *Designerly ways of knowing*. New York, NY: Springer.

Csikszentmihalyi, M. (1991). Design and order in everyday life. *Design Issues*, *8*(1), 26–34.

Csikszentmihalyi, M. (1996). *Creativity, flow and the psychology of discovery and invention*. New York, NY: Harper Perennial.

Dawidowsky, F. (1905). *Glue, gelatine, animal charcoal, phosphorus, cements, pastes, and mucilages* (2nd ed.). Philadelphia, PA: Henry Carey Baird & Co.

Groth, C. (2017). Making sense through hands: design and craft practice analysed as embodied cognition. Doctoral dissertation, Aalto University, Finland.

Häggman, A., Honda, T., & Yang, M. C. (2013). The influence of timing in exploratory prototyping and other activities in design projects. Paper presented at the 25th International Conference on Design Theory and Methodology, ASME 2013 Power Transmission and Gearing Conference, Portland, OR.

Hansen, F. T. (2010). Material-driven 3D digital form giving, experimental use and integration of digital media in the field of ceramics. Doctoral dissertation, the Danish Design School.

Harper, K. (2017). *Aesthetic sustainability* (R. Rahbek Simonsen Trans.). Abingdon-on-Thames, UK: Routledge.

Hemmilä, V., Adamopoulos, S., Karlsson, O., & Kumar, A. (2017). Development of sustainable bio-adhesives for engineered wood panels – a review. *Rsc Advances*, *7*(61), 38,604–38,630.

Hilpinen, R. (2011). Artifact. In *Stanford Encyclopedia of Philosophy*, 11 October. Retrieved from https://plato.stanford.edu/archives/win2011/entries/artifact/.

Houde, S., & Hill, C. (1997). What do prototypes prototype? In M. G. Helander, T. K. Landauer & P. V. Prabhu (Eds.), *Handbook of human–computer interaction* (2nd ed., pp. 367–381). Amsterdam, Netherlands: Elsevier.

Ingold, T. (2013). *Making: anthropology, archaeology, art and architecture*. Abingdon-on-Thames, UK: Taylor & Francis.

Ingold, T., & Hallam, E. (2007). Creativity and cultural improvisation: an introduction. In T. Ingold & E. Hallam (Eds.), *Creativity and cultural improvisation* (pp. 1–23). Oxford, UK: Berg.

Itten, J. (1975). *Design and form: the basic course at the Bauhaus and later.* Hoboken, NJ: John Wiley & Sons.

Karana, E., Hekkert, P., & Kandachar, P. (2010). A tool for meaning driven materials selection. *Materials & Design, 31*(6), 2932–2941.

Kolodner, J. L., & Wills, L. M. (1996). Powers of observation in creative design. *Design Studies, 17*(4), 385–416.

Krogh, P. G., Markussen, T., & Bang, A. L. (2015). Ways of drifting: five methods of experimentation in research through design. In A. Chakrabarti (Ed.), *ICoRD'15: research into design across boundaries, volume 1* (pp. 39–50). New York, NY: Springer.

Lawson, B. (2006). *How designers think: the design process demystified.* Abingdon-on-Thames, UK: Routledge.

Lawson, B. (2012). *What designers know.* Abingdon-on-Thames, UK: Routledge.

Miodownik, M. (2003). The case for teaching the arts. *Materials Today, 6*(12), 36–42.

Miodownik, M. (2005). Facts not opinions? *Nature Materials, 4*(7), 506–508.

Moholy-Nagy, L. (1998/1923). The new typography. In V. Kolocotroni, J. Goldman & O. Taxidou (Eds.), *Modernism: an anthology of sources and documents.* Edinburgh, UK: Edinburgh University Press.

Nimkulrat, N. (2009). Creation of artifacts as a vehicle for design research. Paper presented at the 3rd Nordic Design Research Conference: Engaging Artefacts, Oslo Norway

Nimkulrat, N. (2012). Voice of material in transforming meaning of artefacts. Paper presented at the Design Research Society Conference, Bangkok, Thailand. Retrieved from https://repository.lboro.ac.uk/articles/Voice_of_material_in_transforming_meaning_of_artefacts/9332327.

OED. (2017). OED online. Retrieved 20/07/2017, from http://www.oed.com.

Pedersen, L. T. (2008). *Kreativitet skal læres! når talent bliver til innovation.* Aalborg, Denmark: Aalborg Universitetsforlag.

Romans, M. (2005). *Histories of art and design education: collected essays.* Bristol, UK: Intellect Books.

Ruskin, J. (1971/1857). *The elements of drawing.* New York, NY: Dover Publications.

Schlecht, L., & Yang, M. (2014). Impact of prototyping resource environments and timing of awareness of constraints on idea generation in product design. *Technovation, 34*(4), 223–231.

Schön, D. A. (1983). *The reflective practitioner: how professionals think in action.* Abingdon-on-Thames, UK: Routledge.

Schön, D. A. (1988). Designing: rules, types and words. *Design Studies, 9*(3), 181–190.

Schön, D. A. (1992). Designing as reflective conversation with the materials of a design situation. *Knowledge-Based Systems, 5*(1), 3–14.

Sparre-Petersen, M. (2017). Recycle: about sustainability in glass craft and design. Doctoral dissertation, the Royal Danish Academy of Fine Arts, School of Design.

Suwa, M., & Tversky, B. (1997). What do architects and students perceive in their design sketches? A protocol analysis. *Design Studies*, *18*(4), 385–403.

Tanggaard, L. (2014). A situated model of creative learning. *European Educational Research Journal*, *13*(1), 107–116.

Ullman, D. G. (1992). *The mechanical design process*. New York, NY: McGraw-Hill.

Ullman, D. G., Dietterich, T. G., & Stauffer, L. A. (1988). A model of the mechanical design process based on empirical data. *Ai Edam*, *2*(1), 33–52.

Ullman, D. G., Wood, S., & Craig, D. (1990). The importance of drawing in the mechanical design process. *Computers & Graphics*, *14*(2), 263–274.

Watson, J. (2019). *Lo-tek: Design by radical indigenism*. Cologne, Germany: Taschen.

Wilson, M. (2002). Six views of embodied cognition. *Psychonomic Bulletin & Review*, *9*(4), 625–636.

Yang, M. C. (2004). An examination of prototyping and design outcome. Paper presented at the *ASME Design Engineering Technical Conferences*, Salt Lake City, UT.

Yang, M. C. (2005). A study of prototypes, design activity, and design outcome. *Design Studies*, *26*(6), 649–669.

Courtesy: Copenhagen School of Design and Technology, KEA.

5 Implications for design education

When designers properly study, explore and experiment with a material they will discover details about its source and history, learn about its composition and compatibility with other materials, appreciate its experiential values and achieve an understanding of how to manipulate the material for manufacturing. It is not difficult to recognise the potential this has to provide the designer with the material knowledge required to design for sustainability. However, designing informed and inspired by a material reality, conducting material experiments and using iterative prototyping as a way to develop a design is difficult to learn merely from reading and attending lectures, simply because there is a type of knowledge involved when designing with materials that cannot be learned in this way.

Ironically this also means that whereas this book may be entirely intelligible to designers or design educators with little experience in material experimentation and making, it will not on its own enable them to design with materials. Thus, should a design school wish to introduce a design process that includes a material dialogue to their students, it might in some cases be necessary to demand that the students acquire different skills and knowledge and, depending on the educational practice at the individual design institution, it may require changes to curriculum, physical facilities and not least didactics and therefore potentially take time to implement.

As already described in Chapter 2, the role of materials in the design process and design education has altered through different periods in time and whereas materials and making seem to be on the way back, it will probably take another decade or two before we can accurately conclude if and how this is affecting design education on the whole. Despite general trends that could have an impact on most design institutions, it is not possible to produce one single definition of contemporary design education. Some rest on a strong artistic heritage, others are built on a more technical foundation and some are part of large universities and may have a solid theoretical core but do not necessarily relate to craft practice. These differences will be reflected in the didactic approach, the curriculum, the physical facilities and potentially even the understanding of the activity of design. Consequently, if materials are to be reinstated into design education and returned to a central position in the design process it will likely affect some design educations on several levels, but there is no one answer to the question of what the implications will be. A few educational programmes may already be close to working in this way and to others it would require significant alterations.

Including a material dialogue into the design process simply means introducing a different way of getting from initial idea to finished design. Furthermore, because a design inspired, informed and restricted by a material reality does not eliminate other aspects of importance for the designer, it remains relevant to understand human behaviour, aesthetics, ergonomics, form, drawing, etc. Therefore, the aim of addressing educational practice here is not to propose one complete programme for a new kind of design education.

Nevertheless, it is important to address educational practices because there are certain qualifications which enable the designer to design in with materials and without these prerequisites he or she may struggle to extract information from the material and to conduct the material experimentation and build the prototypes. Thus, in the following discussion of implications for design education the focus is primarily on these prerequisites and how these requirements should be reflected in educational practice.

5.1 Prerequisites

In this section I will address what is required and in the following I will describe the learning theories for how these qualifications and knowledge are obtained and can be taught in design education. The prerequisites are divided into making skills, experimental practice and apt knowledge. Nevertheless, the reason it is important to address the educational practice involved is that there are certain qualifications which enable the designer to design with materials. In this section I will address what is required and in the following section I will describe the learning theories for how these qualifications and knowledge are obtained and can be taught in design education.

5.1.1 Making skills

David Pye has said that there are a sufficient number of definitions for the word 'skill' that one could easily start an argument (Pye, 1968). And his ideas of skill as a purposefully constrained physical action would likely clash with Frayling's later and more inclusive definition of skill as manual dexterity, craft experience, conceptual activity, general know-how or even a shifting combination of these four (Frayling, 2012). Within a design context, skills may be involved in many situations; e.g. when drawing, handling tools, using techniques or software; or even when employing the social skills necessary to communicate well with manufacturers and users. Thus, making skills are far from the only skills relevant to a designer. But they are the kind that will be discussed here, as they are critical in the process of manipulating materials and building prototypes.

If design students have been trained in a design process that is largely immaterial, they may not only lack knowledge of materials, but likely also the skills and techniques required to work with materials. It is relatively easy to detect if students lack making skills: they will tend to be ineffective and cumbersome in their handling of the tools, and if the use of one tool proves unsuccessful, they will not necessarily know which other tools could be utilised instead. When they do not have the experience of manipulating materials and seeing how they can turn into things, they typically possess a very limited embodied knowledge to build upon. Not surprisingly having previous experience of working with a material and a tool will enable and contribute to the student's ability to learn new tools and explore new materials. These experiences also include knowledge obtained from outside their design education: students will often even refer to basic experiences from primary or secondary school, for example working with felting wool or modelling in clay.

Interestingly, students do not always have an accurate perception of their own making skills and ability to experiment with materials. Some students may declare that they have no experience of manipulating materials and using tools, but may in fact have relevant experiences from other fields that they are not aware of, such as cooking, which includes very useful knowledge of how to handle and treat a variety of ingredients, and these can be transferred into the manipulation of bio-materials. Contrary to this, other students can have considerable experience with digital design and manufacturing and therefore classify themselves as makers, but whereas this can be a great asset in the design process and useful for some types of prototyping, it does not necessarily mean that they have any embodied knowledge of materials and how to manipulate them. They may have designed and produced several objects, but the material will frequently have been limited and predefined by the tool. (I will return to how digital manufacturing tools, such as a 3D printers, can be used in a material dialogue and also discuss the correlation between having simple traditional craft skills and the ability to understand more advanced digital manufacturing processes later in this chapter.)

The most conspicuous problem regarding poor making skills is low-quality material samples and prototypes that lack in finish and fabrication. However, less visibly but perhaps more importantly, poor making skills will also affect the information derived from the material dialogue. A richer and more varied material exploration will provide the designer with ample information for inspiration and a more detailed knowledge about the material. Likewise, a series of well-executed prototypes will not just ensure a more tested and debugged design, but also provide the designer with more precise information during the process for the product development.

As noted before, designers are unlikely to reach the same level of skills as a craftsman, but basic making skills are essential for manipulating and exploring materials – even if some of the prototypes will be made digitally and the final product will be fabricated without the need for manually handling the material. Basic making skills will enable the designer to achieve a certain level of quality in material experimentation and prototyping, and this is essential for extracting information and inspiration from the design process. Still, to explore and develop a material, as well as to learn from the iterative prototyping of a design, requires more than the ability to manipulate materials and handle tools. It demands a system, the ability to ask the right questions and to set up/build, conduct, document, evaluate and learn from an experiment/a prototype.

5.1.2 Experimental practice

Experiments have played an important role in the history of both the sciences and the arts, but they rest on fundamentally different traditions and practices. The artistic experiment has historically been closely linked to the avant-garde and has been a way to break with tradition. It characteristically involves a process that can enable the emergence of the unforeseen and dissolve boundaries. Experiments in the arts do not have a set of specific rules or procedures that must be adhered to, but typically implicate exploring something new, e.g. by applying an unusual technique, a different method, a new material or composition.

Contrary to this, a classic scientific experiment is founded on systematic enquiry and typically concerned with testing a hypothesis or a known fact based on existing theory. It is controlled, follows a predefined protocol, is meticulously documented and should be repeatable by other scientists. Unsurprisingly the two approaches to experimentation result in radically different 'products'. The scientific experiment will provide a result from which quantifiable data can be extracted and analysed; the outcome of the artistic experiment may equally provide valuable knowledge or contributions, but it is rarely quantifiable and, thus, must be understood in a different way.

Former director of the Max Planck Institute Hans-jörg Rheinberger distinguishes between these two types of results as 'technical objects' and 'epistemic things'. He defines technical objects as the products of an investigation set up to provide answers, whereas epistemic things embody what one does not yet know and are therefore more likely to generate new questions. He describes 'experimental systems' as the centre and the vehicle of modern scientific research, but he also points out that in actual fact the sharp division between a strict scientific approach and the more explorative artistic experimentation does not always represent the reality of experimental practice, nor is it necessarily desirable. He describes how scientific experiments are not merely methodological vehicles to test knowledge that has already been theoretically grounded or hypothetically postulated, but experiments must also be the actual generators of knowledge and in order to be this, the experimental system needs to be sufficiently open to produce unforeseen results and let new technologies and instruments in.

Relatedly, Schön argues that in practice there are several other ways to experiment than a strictly scientific approach and highlights what he calls problem-setting experiments, where the problem or the question is set in a way that it can be solved through experimentation; exploratory experiments, where action is undertaken to see what follows; and move-testing experiments, where action is undertaken to achieve an intended change. Schön stresses how design problems can be ill-suited for technical rationality and describes how the practitioner may use experimentation with the aim of improving things or transforming a situation. Consequently, the knowledge extracted from the experimentation is not the ultimate goal, but rather a means for successful intervention (Schön, 1983).

Interestingly, experiments in practice-led design research are often placed in between the scientific and the artistic tradition (Steffen, 2013) and may therefore contribute with both technical objects and epistemic things. Experimentation is a recurrent central driver for practice-based design research and is gradually becoming more well-defined (Bang & Eriksen, 2014; Brandt & Binder, 2007; Krogh, Markussen & Bang, 2015). Koskinen et al. stress that it is the theoretical scaffolding that makes an experiment in design research different from an experiment in design practice (Koskinen, Binder & Redström, 2008), but despite the fact that the words 'experiment' and 'experimental' are frequently used in design, the role and definition of experiments in design practice are often somewhat blurry (Steffen, 2013).

In a design process that includes a material dialogue, exploring and developing both the material and the design are grounded in experimental practice; this means that the designer will benefit from having a certain level of theorical understanding of experimental systems, but most importantly the designer needs to know how to practise. This requires the ability to set up, conduct, document and evaluate an experiment according to the type of enquiry. Like experimentation in design research, the experimentation in a design process that includes a material dialogue frequently reaches into both an artistic and a scientific tradition and therefore also produces both technical objects and epistemic things. And sometimes the result of an experiment embodies both. Rheinberger describes this as a functional relationship between them when epistemic things have reached a certain point of clarification and can be transformed into technical objects and vice versa (Rheinberger, 2013).

By providing designers with a piece of wood and a knife, asking them to make a spoon, they will almost automatically enter into a material dialogue (Bak-Andersen, 2019). However, solving a design problem through a design process that includes a material dialogue is significantly more complex and it is entirely up to the designer to construct a suitable experimental set-up for material experimentation and prototyping. This entails experimental practice and it is very noticeable when design students start working in the lab or workshop without this.

Design students who engage in a design process that includes a material dialogue without any knowledge or experience of experimental practice will tend to have a random explorative approach and therefore use whatever tool is at hand, without considering its suitability. They are often unsystematic and inefficient, which may result in issues such as their process of experimentation stalling, because they have to wait for the results from a single experiment before being able to make the next, when they could have made several variations of the same experiment simultaneously and reached an answer much faster. They will tend to be imprecise in their documentation (or even refrain from documenting their process) and therefore be unable to repeat a successful result, and, finally, they may struggle to extract and use the knowledge from their results – whether they are quantifiable and suitable for comparable analysis or of an epistemic quality and therefore better extracted by means of reflective practice as defined by Schön (Schön, 1983).

In his book *The reflective practitioner*, Schön writes about two different types of reflective practice: 'reflection-*on*-action' and 'reflection-*in*-action'. 'Reflection-on-action' means to reflect on results and actions after they have taken place, in order to learn and subsequently form the enquiry for the next experiment. He describes 'reflection-in-action' as reflecting on the situation while changes can still be made to affect the outcome, rather than waiting until a later time to reflect on how things could be different in the future (Schön, 1983). Students will typically be able to discuss and rationally justify the decisions they made fully consciously after each design experiment as a result of reflection-*on*-action. However, some of their decisions will likely be based upon reflection-*in*-action and will have taken place simultaneously with the experimentation. Reflection-in-action can be harder for the educator to detect, as it is less explicit. It can be based upon a hunch or intuition and lead to perfectly sensible decisions, but it can also be caused by of fear of failure, or a fear of exposing lack of knowledge or poor making skills.

Thus, design students will need to learn how to set up and conduct different types of experiments and need to understand the reasons for conducting scientific experiments. They must learn to be systematic and to document and register their findings in order to be able to repeat an experiment, even when they are working with a more explorative and artistic approach. Nevertheless, they will need guidance not just on the specifics and technicalities, but also to some degree on a more emotional level. Because even though the artefacts that are produced are typically good telltales for the process and valuable as epistemic artefacts for reflection-on-action, they are rarely all aesthetically pleasing. It requires some confidence as a designer to stay in a process where the material that one is working with for a while is likely to be unappealing in some way. Furthermore, if a student is used to working digitally, the different aesthetics and lack of cleanliness in the process of material experimentation can be distressing. Consequently, an un-guided inexperienced designer may fixate on the first aesthetically pleasing sample encountered, instead of pursuing further experimentation.

Figures 4.2a, 4.3a and 4.4a show the results of material experimentation by students that draws on both a scientific and an artistic approach – a similar approach to material experimentation can be seen in the work of the designers Edvard, Klarenbeek and Dros (Figures 4.8–4.15). Working with experimental practice will also require more concrete things, such as the ability to operate the equipment, tools and machines available and knowing how to work in a sterile environment, and even just keeping a lab journal can make a notable difference. Curiously, as can be seen in Figure 5.1, a lab journal may look very particular in the hands of a design student, and be used not just as a tool for planning and registration but also may include material studies and still perform as a reflective tool, in the same vein as a traditional designer's sketchbook.

Developing a design through material experimentation and iterative prototyping involves an experimental practice that relates to both the scientific and the artistic traditions. In an interview with Michael Schwab, Rheinberger describes how practice will lead to expertise and a deeper understanding of experimentation that may include what he calls *experimental spirit*. In many ways this describes the quintessence of experimental practice in a design process that includes a material dialogue:

> One usually associates 'spirit' with spirituality, a purely mental activity. However, in my understanding 'experimental spirit', the interaction of the experimenter with his or her *material*, lies at the centre. If one is not immersed in, even overwhelmed by, the material with which one works in an experiment, the material itself somehow comes alive. It develops and agency that turns the interaction into a veritable two-way exchange. It's both a forming process and a process of being informed. The experimental spirit has a haptic quality. 'Haptic' here points beyond mere sensory impression; it carries an epistemic connotation.
>
> (Rheinberger, 2013, p. 198)

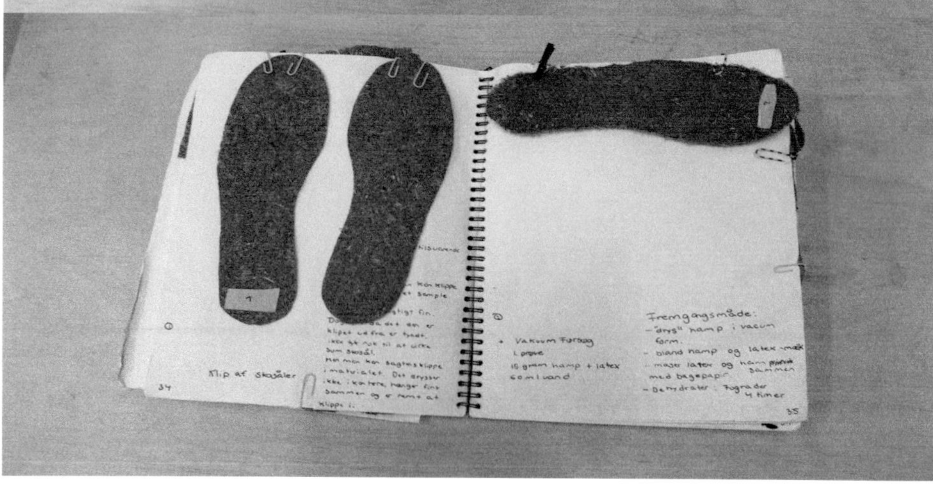

Figure 5.1 **Student's lab journal**

Learning to document process and results supports the student in a more systematic and scientific approach to material experimentation. Interestingly, the student has included material samples and elements from the process, which makes the recording into a hybrid between standard lab journal and a designer's sketchbook. *Courtesy*: Copenhagen Material Design Lab, Copenhagen School of Design and Technology.

5.1.3 Between art, technology and natural sciences

Over the last century, the complexity and amount of materials available to the designer have vastly expanded (Ashby & Johnson, 2013). Designers must relate to new materials and advanced material technology coming from fields which traditionally have not been connected to product design, and as a result of the environmental crisis designers have to use alternative resources and recycled materials. Consequently, when materials are reinstated into a central position within design education, it cannot simply be done in the same way as at the Bauhaus 100 years ago; defining the workshops by the same material categories would be unproductive, and the making skills and understanding of manufacturing processes must reflect today's technological level.

The environmental situation and awareness have changed and, therefore, we can no longer mix materials in any way we like with the sole purpose of achieving products with good mechanical properties and appealing aesthetics that meet the right price point. We must relate to such aspects as the recyclability, biodegradability and toxicity as well as the use of new alternative resources. To do this, the designer needs a basic knowledge of the science behind the material as well as a fundamental understanding of the chemical elements that constitute the material.

The designer must be able to study the material in more detail than can be observed by the naked eye and thus operate not just on the macro-scale, but also to some degree on the micro-scale. When designers experiment and design with materials, they have to relate to knowledge and methods of investigation from natural science. This is particularly evident when using new unexplored or non-standard material. For example, when Edvard grows a mycelium material into a form, he will have to study and understand the biological processes involved to work with the material creatively.

When designing for a circular economy using biomass materials, the designer is rapidly confronted with a variety of incompatible binders and finishes. There is no handbook on how to deal with this and therefore it can be relevant to study nature's way of designing (Figure 5.2). This is referred to as biomimicry (Baumeister, Tocke, Dwyer, Ritter & Benyus, 2014), and when biomimicry is used for more than mimicking structures and concepts and actually involves learning from nature's recipes and survival strategies, it frequently demands a basic knowledge of biology and biochemistry. Knowledge from natural science will likewise be required if designers work with materials that have only recently come out of labs. For example, in order to explore and use a synthetically grown material, such as the honeybee silk shown in Figure 5.3, which is a material that has been cultivated in the labs of the Commonwealth Scientific and Research Organisation

urals
WOOD
IDENTIFICATION & USE

Figure 5.2 **Learning from nature**

Studying the survival strategies of plants can provide answers to how we can waterproof and colour our products in a more sustainable way. Equally, structures built by other animals, such as the wasp nest exhibited in the Box at the Material Design Lab, will provide information about optimised lightweight structures and material design.
Courtesy: Copenhagen Material Design Lab, Copenhagen School of Design and Technology.

(CSIRO) led by the scientist Tara Sutherland (Sutherland, Young, Weisman, Hayashi & Merritt, 2010), the designer must have a basic understanding of silk proteins and synthetic biology.

The designer should not aspire to become a material scientist or a biochemist. However, a design process that includes a material dialogue will inevitably extend into different fields of natural science, such as material science, chemistry and even biology and biotechnology. The designer does not need extensive expertise in these but must have sufficient knowledge to be able to talk to, and work with, a broad range of specialists in the natural sciences. This is easier said than done, because what is at stake is not just formal theoretical knowledge, such as data on materials, chemical formulas and processes, which in principle can be learned from reading existing literature. Applying it in a design process *also* requires practical skills: comparing data on materials may require the ability to operate the tensile strength machine, working with mycelium would oblige the designer to know how to sterilise the working environment, understanding processes at the micro-level will entail using a microscope, etc. These are not necessarily skills that current students will acquire during their design education; they are difficult to learn simply by reading and not necessarily a type of expertise that can be found within the school's existing faculty either.

Most design students I have met appear to have an aptitude for understanding aspects related to the social sciences, which enables them to do historical research on a material or to understand the concept of the cultural meaning of a material and how this might be transferred into a product. In contrast, acquiring knowledge from the natural sciences often seems an insurmountable obstacle for them. This may be a result of their general education, or their design education, which in some places includes both sociologists and anthropologists in the faculty, but it could also be due to our experiential knowledge from life in general. Steen Nepper Larsen, a Danish philosopher, writes about how the human scale of experience has another wavelength than biochemistry. He suggests that we learn about the world from living among everyday macroscopic phenomena and know 'objects' such as neighbours, bicycles, coffee cups and books, but we do not automatically learn about the world at a micro-level. We do not experience the level of blood cells, grasp the immune system or feel neuronal activities (Larsen, 2010).

Knowledge from social sciences, such as anthropology, psychology or economics, is important for designers, but a design process which includes a material dialogue is likely to be more closely connected to the natural sciences than more immaterial design processes. This pushes the activity of designing physical objects towards a field defined by art, technology and the natural sciences. This

Figure 5.3 **Synthetically grown honeybee silk in front of honeycomb from a beehive**
Like many other insects, the honeybee produces a silk with unique properties. Approximately 30% of the natural honeycomb made by the bee is silk and it works as a reinforcement of the wax in the construction. The protein solution for the wet-spun silk yarn in front of the honeycomb in the picture was grown by means of synthetic biology by a team of scientists at the CSIRO in Australia led by Tara Sutherland.

connection sparks more general considerations regarding the field of design, but it also reveals very interesting ground for further research, because the ability to work at this intersection could potentially provide designers with a better understanding of the physical realm of the products that they design, which would prepare them for communicating with scientists and for applying new emerging materials creatively. This connection to natural science should ideally be reflected in the curriculum, and, as I will address in the following, it should also be visible in both the workshop facilities and the faculty.

5.1.4 The physical facilities

Judging from the website presentations of design schools, it would appear that most institutions that offer product design courses and the like have some degree of workshop facilities. Some schools call their educational spaces that are adapted to work with materials and making 'workshops', while others describe them as 'laboratories' or simply 'labs'. The physical spaces that can be seen as part of the Maker Movement are typically called labs – Makerlabs, Fablabs, Biolabs, etc. Naturally a title does not by default define the activity within the space, but it is worth noting that the meaning of the word 'workshop' indicates a space for making and manufacturing, whereas 'laboratory' suggests a space for experimental practice or testing.

Whether the physical facilities are suitable for working with materials and making in the way required by a design process that includes a material dialogue primarily depends on two aspects: the set-up of the facilities, defined by the available tools, installations and equipment; and the didactic approach, defined by the curriculum and the faculty in charge. If a workshop is set up to make prototypes with general woodworking tools, it is not necessarily suitable for material experimentation. Simply because it is likely to primarily be prepared for working with form and will consequently be ill-suited for working with the interface between materials research and design, which must support activities such as melting, mixing, growing, shredding, testing, forming, pressing, spinning, knitting and weaving. In the article 'Design tools for the interdisciplinary translation of material experiences' the authors describe how the right facilities and tools with enable designers to influence the development of a material at the stage that it emerges from the laboratory (Wilkes et al., 2016).

Secondly, the suitability depends on the practice that defines the space. In a design process that is primarily conceptual and immaterial, the purpose of building a prototype will typically be to visualise and present a final design idea, but in a design process that includes a material dialogue iterative prototyping is a vehicle for developing the design. From a practical point of view this means that most of the design process will need to take place in the laboratory/workshop. However, to learn to design with materials also requires the support of a faculty member, who will not just assist the student with technical issues related to making but have the expertise to support the student in the experimental practice involved in conducting experiments and building prototypes.

There is no exact recipe for this kind of space and ideally it will be a combination of different labs and workshop facilities. Some making activities cannot be combined, e.g. there are very good practical and safety reasons why a metal workshop and wood workshop should be kept separate. It would obviously be reckless to have a furnace for melting metal close to an extraction filled with sawdust and equally the sawdust can destroy the filters in the extraction on a fume cupboard in a laboratory. Furthermore, the profile and the educational programmes of each individual institution will influence the set-up of a workspace designed to work with materials. To exemplify the diversity of spaces set up to work with experimental practice, materials and making, I will introduce a few different places that are set up for working with materials and making within art, design and architecture. They vary in their set-up, but share a playful, hands-on approach to material experimentation and making.

The Material Design Lab is a cross-disciplinary facility designed to work with materials at the intersection between art, technology and natural science. It is located at Copenhagen School of Design and Technology and supports all educational programmes related to materials and making. The Material Design Lab was designed and developed with the same objective and motivation as this book, and the results from the activities in the lab have informed, and been informed by, my research simultaneously in the period from 2013 to 2018. Even so, the laboratory at the Material Design Lab should not be understood as a conclusive solution for a space suitable for designing with materials, but rather as a prototype from which it is possible to learn and develop. The Material Design Lab was first presented in 2013 (Bak-Andersen, 2013) and then officially opened in January 2015. Over the years the lab has developed – not just the physical facilities but also the theoretical foundations of the lab. The comprehension of the constituents of material experimentation, the didactic approach and the understanding of creative learning environments have in this period been expanded and substantiated. At present, the Material Design Lab consists of a Material Connexion materials library with commercially available materials (Figure 5.4), a separate exhibition of raw materials and narratives about materials (Figures 5.5a and 5.5b), and finally the laboratory, which is the primary focus in this context (Figures 5.6–5.8b).

The cross-disciplinary properties of the lab are visible in the installations and equipment in the space. As can be seen in the Figures 5.6–5.8b, the appearance is perhaps best described as a mixture between scientific laboratory and an industrial kitchen with elements from a more typical design prototyping workshop. The lab features a countertop with a large work area, hotplates, sinks, a washing machine and dryer, two large ovens, an extraction system, a dehydrator and cupboards full of pots, pans, blenders and cooking utensils – all of which could be found in an industrial kitchen. But the lab also contains a fume cupboard, different microscopes,

Figure 5.4 **The Materials Library at the Material Design Lab at Copenhagen School of Design and Technology**
The 1,500 materials displayed in this library are all commercially available and supplied by Material Connexion (www.materialconnexion.com). Each material is accompanied by a short description, but detailed information and many more materials can be found in the material library's database.
Courtesy: Copenhagen Material Design Lab, Copenhagen School of Design and Technology.

(a)

(b)

Figures 5.5a and 5.5b **The Box at the Material Design Lab at Copenhagen School of Design and Technology.** 'The Box' is a collection of raw and mono materials at the Material Design Lab at Copenhagen School of Design and Technology. It includes a variety of sample materials before they are mixed, processed or treated, examples of materials and constructions made by animals (see Figure 5.2), a collection of natural finishes, adhesives and dyes and examples of material processes, e.g. the steps in recycling a plastic bottle into a fleece fabric.
Courtesy: Copenhagen Material Design Lab, Copenhagen School of Design and Technology.

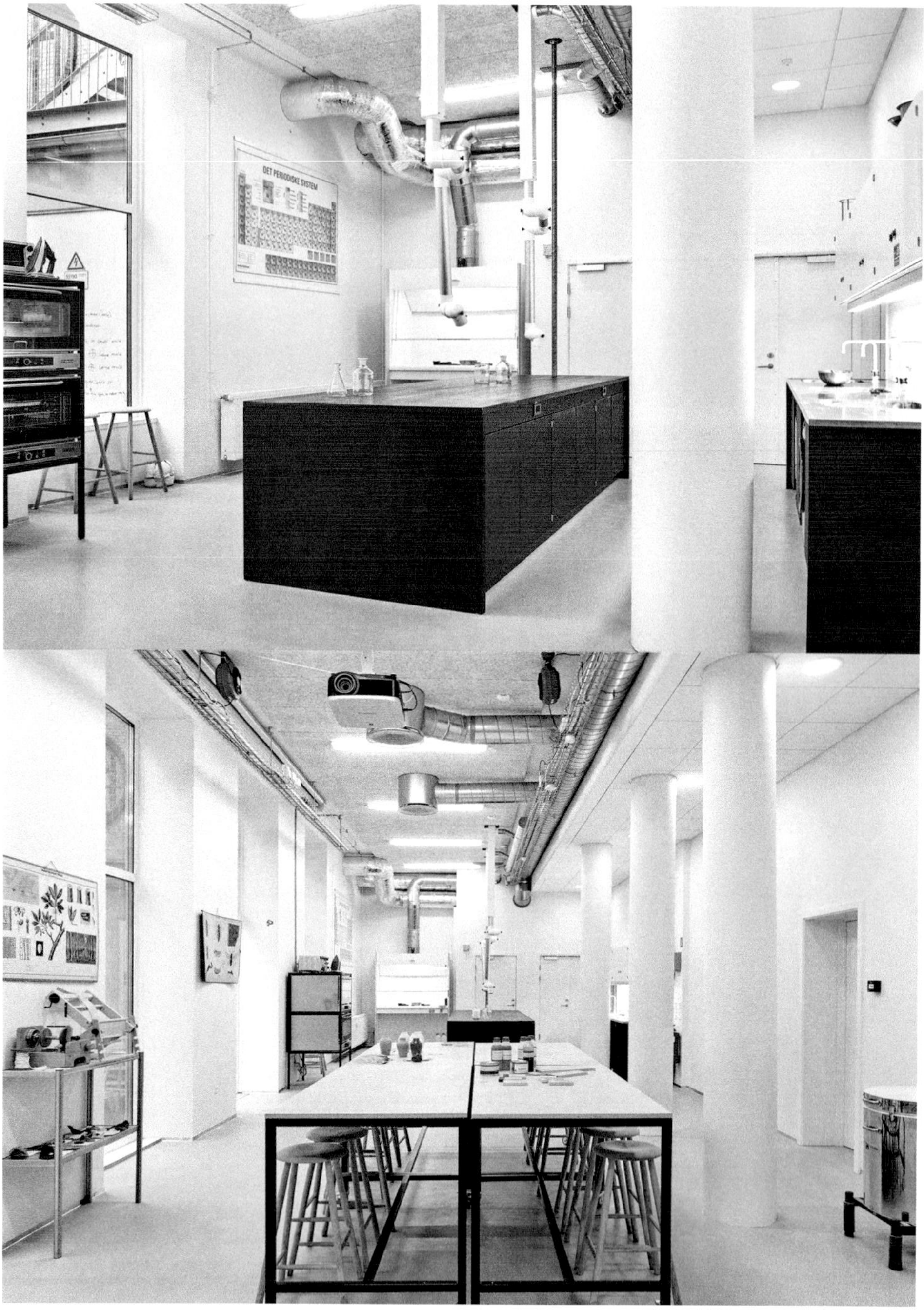

Figure 5.6 **The lab at the Material Design Lab at Copenhagen School of Design and Technology**

The lab was designed on the bases of research on material experimentation and the experiences from professional and educational projects exploring ways to design with materials. Since the opening it has expanded, been adapted, new tools have been built and added and, hopefully, it will continue to develop in the coming years.

Courtesy: Copenhagen Material Design Lab, Copenhagen School of Design and Technology.

Figure 5.7 **The lab at the Material Design Lab at Copenhagen School of Design and Technology.**
Frequently the students are asked to work on more unusual materials. Here they are (in the photograph at the top) presented with: apple pulp, a waste material from apple cider and apple juice production; unwashed wool, a waste material from meat sheep farming; egg shells, a common waste material from food production; offcuts from straw from old thatched roofs (when thatched roofs are replaced with new, the old straw is still usable for other purposes than thatching) and flax fibre production. The bottom photograph shows sugar beet pulp, which is a waste material from the production of sugar. By asking students to experiment with lesser-known materials, they engage in a more open-minded exploration with fewer references to industry standards. They are frequently forced to seek knowledge about the materials outside the field of design, which may involve collaborating with companies and experts outside the school. By working with these kinds of materials, the students learn that there is a much broader range of materials with excellent potential and properties that can be used for design, but which may not be available in the shops where they would normally buy their materials.
Courtesy: Copenhagen Material Design Lab, Copenhagen School of Design and Technology.

(a)

(b)

Figures 5.8a and 5.8b **The lab at the Material Design Lab at Copenhagen School of Design and Technology**

As can be seen, material experimentation includes many different activities. Designing with materials is considerably messier than digital or conceptual design; it requires more tools and equipment, installations such as taps, sinks, floor drains, a solid extraction and ventilation system – and it requires space. The total area of the lab is approximately 100m^2 and it is suitable for a maximum of 20 students; this is perhaps a detail, but it is important to consider if a school wishes to integrate the use of laboratories and workshops in an educational programme. *Courtesy*: Copenhagen Material Design Lab, Copenhagen School of Design and Technology.

measuring equipment and lockers with Bunsen burners, Petri dishes, beakers, test tubes and pipettes, which makes the space resemble a scientific laboratory. Finally, there is such equipment as a vacuum former, a ceramic kiln as well as tools like carts, a spinning wheel and a small space for mould making, elements that would more frequently be found in a design prototyping workshop. The original intention was to create a laboratory for exploring materials on both micro- and macro-scales. To a large extent this has been accomplished. However, the ambition was also to abandon traditional material categories for workshops. This has been done as far as possible, but it has not been feasible to make one space that is equally suitable for exploring all types of materials. The lab is at present well equipped for exploring biomaterials, but, due to limited space and other practical reasons, it has not been possible to include, e.g., proper tools and equipment for exploring metals.

The lab was initially developed in a controlled and planned manner, but in the later years it has been a more organic process, where the space has adapted and responded to the activities and demands of the many different people using the space. E.g. faculty from the Production Technology programme made it clear that they needed more data on the materials they were working with and as a result a tensile strength machine was installed and access to CES EduPack, a database with resources for materials education, was provided (Ashby, 2008). Student-led projects have added to the equipment, an example of which is a plastic shredder and extruder built by students who needed to experiment with the recycling of plastics. Similarly, obstacles such as the inflexibility of certain tools for material exploration have provoked progress, e.g. frustration with standard 3D filament printers which pushed forward the build of a new 3D printer adapted to work with semi-solids and support experimentation with biomaterials in additive manufacturing.

Reinstating working with materials into design education may require a different type of physical facility than is at present available at most design schools. But the Material Design Lab is not necessarily ideal for all activities in a design process that includes a material dialogue. In general, it is better equipped for studying, exploring and experimenting with materials and less suitable for iterative prototyping, which is likely to be more concerned with activities such as forming, cutting, sewing, assembling and printing. Some of this can be done in the lab, but most likely it will at some point be necessary to use the tools in other workshops. The equipment in the lab is primarily focused on learning and hence the tools are deliberately of a fairly rudimentary design and technological level, simply because they need to be relatively easy to handle in order for the students to be able to 'tinker' and use them creatively. Nevertheless, this also means it has been beneficial to collaborate with the Danish Technological Institute, in order to support students who have developed materials and prototypes suitable for commercial industrial production. An example of this is the apple leather developed by Hannah Michaud presented in Chapter 4 (Figure 4.1). In larger educational institutions, a higher level of technical support with more advanced labs and manufacturing equipment may be available inhouse. From a practical point of view, the lab is relatively small, which means that one project or course easily occupies the entire lab for a period of time and makes it impossible to work with other materials and processes at the same time – partly because of a restriction in capacity, but also because one material experiment may pollute the other if conducted in proximity. Of course, it is inspiring to see how one week the lab can be dedicated to working with natural dyes on textiles and the following weeks is transformed into a space for growing mycelium. However, it is also a good indication of how a design process that includes a material dialogue is rather demanding in terms of space, equipment and tools.

The Institute of Advanced Architecture of Catalonia, IAAC, in Barcelona, has a number of Master's degree programmes focused on digital architecture and fabrication. IAAC was the first institution to open a Fab Lab in Europe in 2007 and it is by now a place where the Maker Movement is fully integrated. The institute has an experimental and productive approach to architecture and this approach, as well as the emphasis on digital fabrication and technology, are clearly reflected in the spaces set up to work with making and materials. The institute is located in an old factory and the workspaces include digital fabrication tools such as a variety of 3D printers, CNC machines and robotics arms (Figure 5.9). The institute does have a minor space specifically dedicated to material experimentation, but in general the workspaces are large and flexible in their set-up. One of the institute's more unusual labs is on the Valldaura campus. Valldaura is a 19th-century farmhouse in the countryside outside the city. Students and researchers live and work there in a community, managing the forest on the large estate and producing food. The location of the lab provides an innate connection to natural materials and local resources (see Figure 5.10) and the spacious surroundings allow for developing large full-scale prototypes.

Many of the educational activities at IAAC are placed in a field between art, technology and natural science. The Master's degree programme, 'Design for Emergent Futures', is a good example of this. It is a collaboration between IAAC and the Fab Academy, a digital fabrication programme directed by Neil Gershenfeld of MIT's 'Center for Bits and Atoms', which is based upon MIT's rapid prototyping course: 'How to Make (Almost) Anything' (Fab Foundation, 2018). Interestingly, some of the faculty at IAAC are not just focused on how to make (almost) anything, but have also been involved with the 'Bio Academy' – 'How to *grow* (almost) anything' – a synthetic biology programme directed by George Church, Professor of Genetics at Harvard Medical School (Church, 2018). Some of the experiences from the Bio Academy were carried forward by the faculty into the Master's programme as a course called 'Biology Zero'. It had the aim of providing the students with a basic understanding of biochemistry, microbiology, cellular biology and synthetic biology (see Figure 5.11).

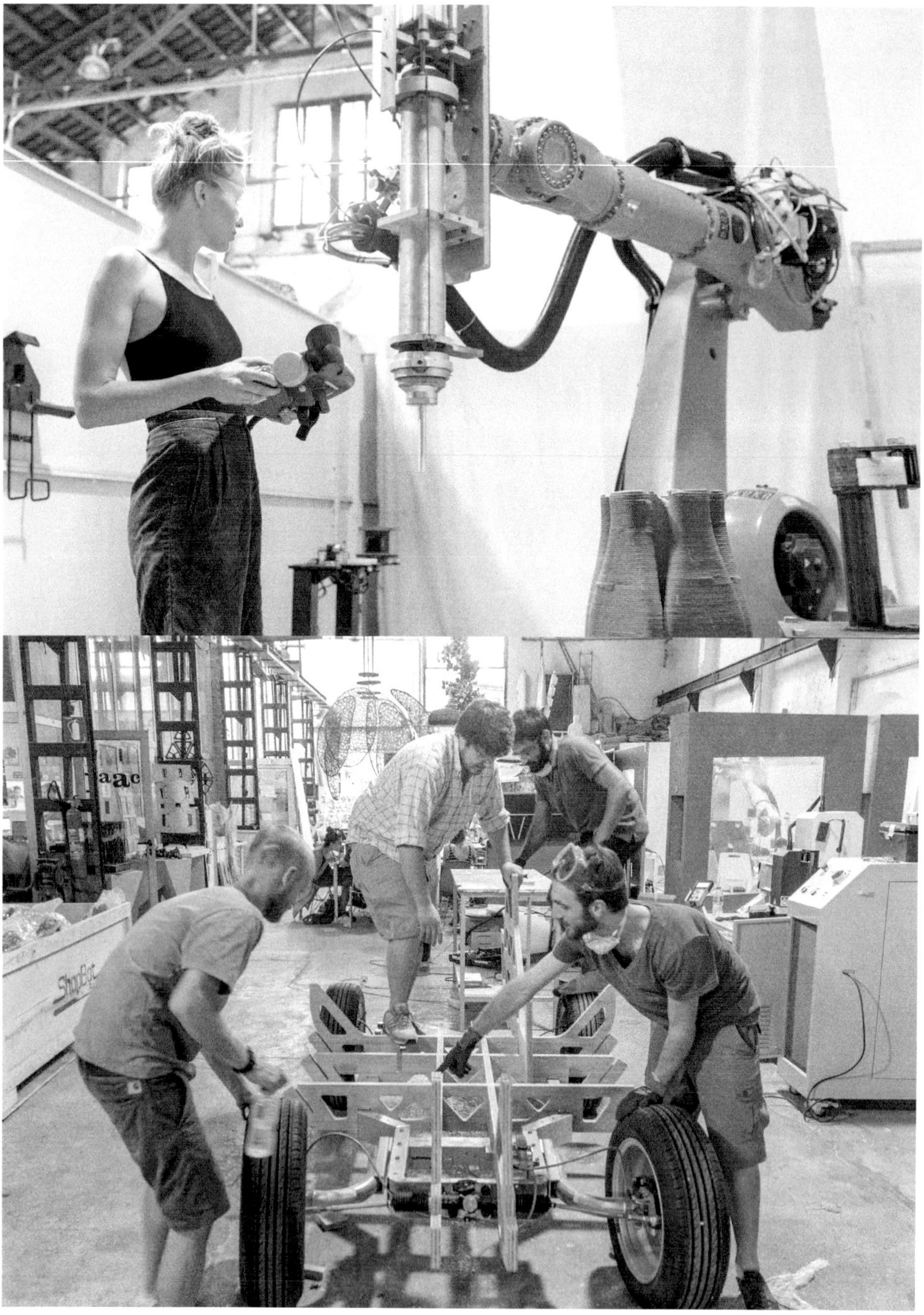

Figure 5.9 **Working at the IAAC workshops**

Naturally, some of the heavier machines are not moved around, but the workspaces for experimenting with fabrication technology and building prototypes are in general not defined to suit one specific type of activity, but to adapt according to the ongoing research, projects and educational activities.

Figure 5.10 **Valldaura**

The estate that surrounds the Valldaura farmhouse offers plenty of resources, and, thus, the opportunity to be self-sufficient and produce fully traceable wood. Many of the educational activities and projects at Valldaura are based on disciplines such as biology, forestry, digital manufacturing and agroecology, and explore the relationship between digital and biological manufacturing.

ROBOTS

Figure 5.11 **The Biology Zero course.**
Using microscopes, incubators and pipettes requires practice and frequently even the types of fine motor skills utilised when handling different tools for craft. In this image, the scientist Nuria Conde Pueyo is teaching the Biology Zero course. She adheres to a strict scientific approach in her work, but at the same time her professional experience has provided her with an embodied knowledge of her field, which, e.g., enables her to determine the status of a sample by smell alone or the ability to feel when the temperature is right for a certain process. Her competencies as a contributory expert enable her to transfer both practical and theoretical knowledge to the students.

CHEMARTS, at the Aalto University in Finland, likewise bridges the gap between art and science. But it is a quite different place from IAAC. CHEMARTS is a collaboration between two schools: the School of Chemical Engineering (CHEM) and the School of Arts, Design and Architecture (ARTS). The idea of the collaboration is to conduct research into the performance and design of advanced cellulosic materials with the objective of inventing new ways to harness wood and cellulose and proposing new ways to use them. What makes the place special is that CHEMARTS students and researchers from chemistry, art, design and architecture explore the potential of the materials together and based on their results create new concepts for the future use of cellulose and other biomaterials (Figures 5.12 and 5.13). The collaboration is defined by the specific disciplines of the partners and the particular focus on biomaterials, but the cross-disciplinary framework is detectable in the physical set-up, the faculty and also the experimental practice. The physical installations and equipment in the lab make it possible to comprehensively test and explore biomaterials at micro- and macro-scales, and by drawing on both an artistic and a traditional scientific approach to experimentation they allow for a design process that includes both systematic testing and creative exploration.

The CHEMARTS cookbook describes working in the interdisciplinary field between chemistry and art and shares methods, projects, suggestions for suitable lab equipment and even recipes. Designers tend to work alone or stay within the designer community, but the authors highlight how designers in collaboration with scientists have the potential to apply groundbreaking findings from scientific material research in everyday life. At CHEMARTS they have observed how design students can make scientifically important observations, because they may experiment on something that the scientists do not consider interesting (Kääriäinen, Tervinen, Vuorinen & Riutta, 2020).

The Institute of Making at University College London is an example of a well-established place which encourages play, research and development of materials and processes. It is set up to support both academic research and hands-on experience. The institute comprises an imaginatively curated material library that frames a multi-purpose space where both hands-on workshops and lectures take place. The front of the space opens up towards the street, which is frequently used for outdoor workshops and masterclasses. In the back there is a well-equipped makerspace set up to support a variety of making activities (see Figures 5.14 and 5.15).

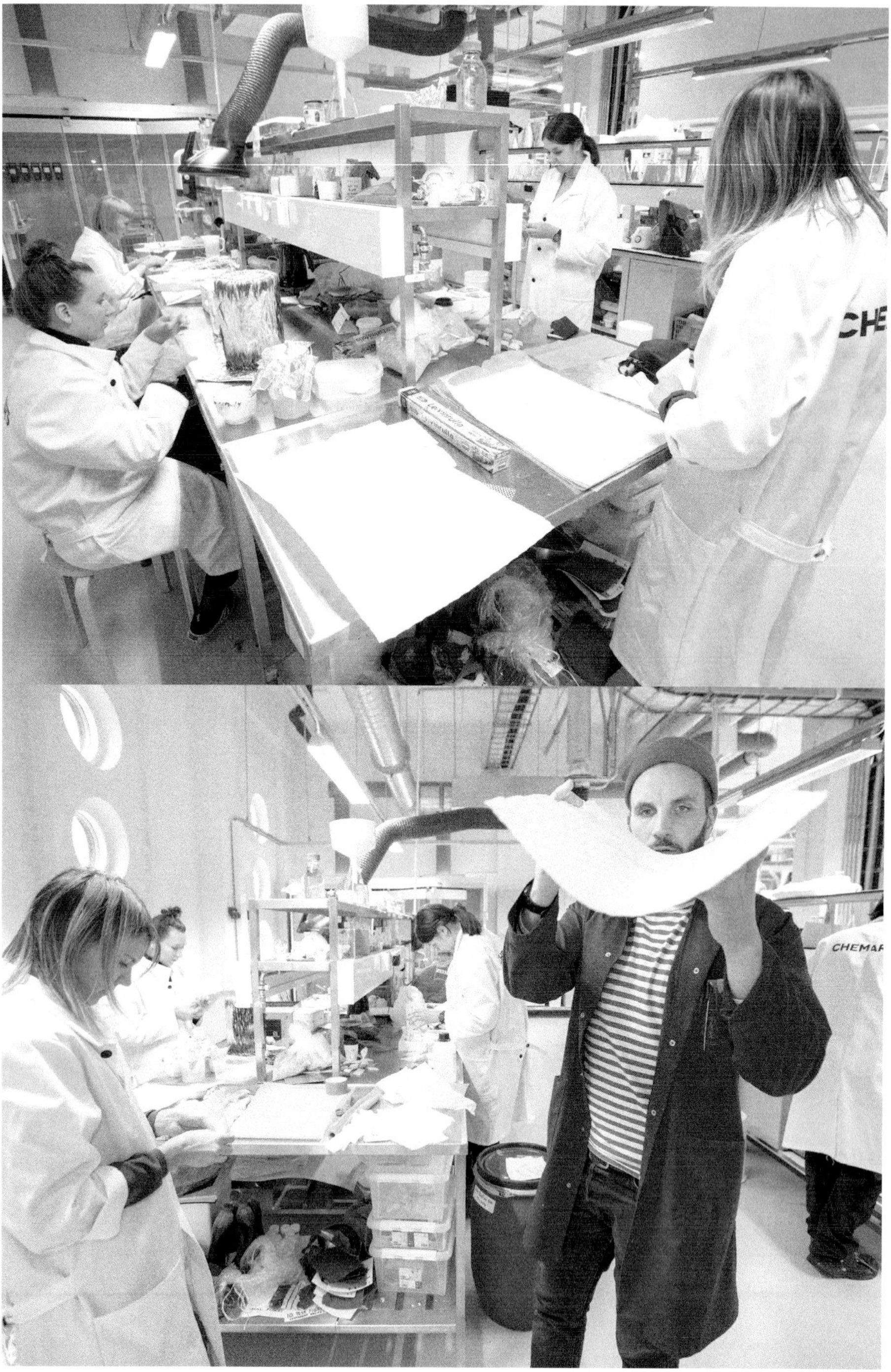

(a)

(b)

(c)

Figures 5.12a–5.12c **CHEMARTS laboratories at Aalto University, Finland**

These images illustrate the collaboration between design and chemistry – two disciplines that are both concerned with matter. It is noticeable how the activities, the experimentation and the material alter in character, scale and expression when designers start working in the chemist's laboratory.

Courtesy: CHEMARTS, Aalto University; photograph: Eeva Suorlahti.

Figure 5.13 **CHEMARTS laboratories at Aalto University, Finland**

Presenting and discussing different material samples and prototypes made in the CHEMARTS workshop.

Courtesy: CHEMARTS, Aalto University; photograph: Eeva Suorlahti.

(a)

(b)

Figures 5.14a and 5.14b **The Institute of Making**

Instead of being positioned as a separate exhibition, the material library is on the walls in the central space of the institute and thus naturally plays a significant role in the activities. The material library includes several curious and inspiring samples. One of the more curated parts is a collection of evenly sized cubes made out of different materials, which provide a unique way of comparing the sensorial aspects of the materials. *Courtesy*: The Institute of Making.

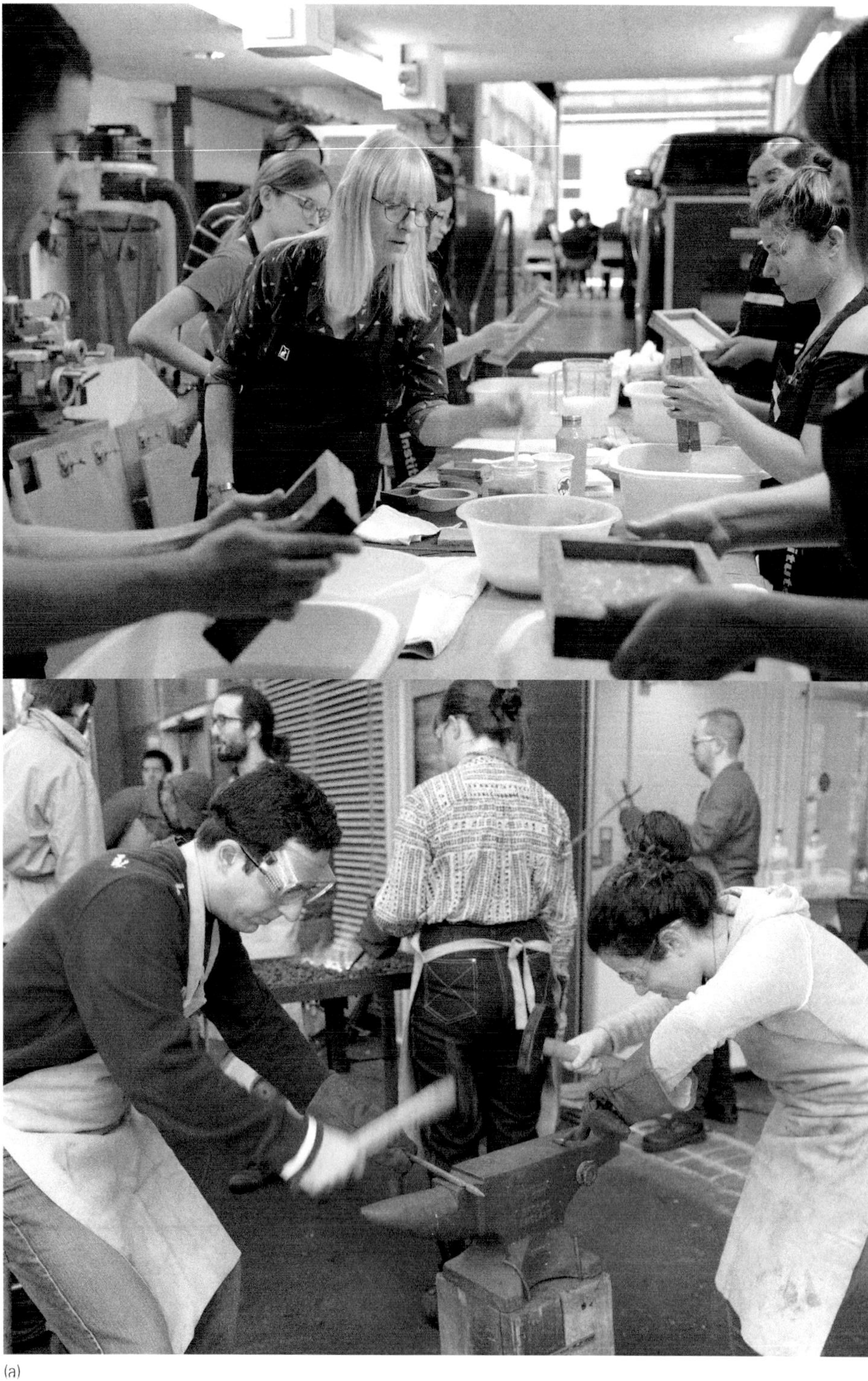

(a)

(b)

Figures 5.15a and 5.15b **Workshops and masterclasses.**
The Institute of Making is focused on all types of materials, making and processes. Their broad range of workshops and masterclasses include everything from microwave steam bending wood to the science of cooking and sand casting with silver. The top image in 5.15a shows a paper-making master class and the bottom is a blacksmithing class. Figure 5.15b is a masterclass in making copper spoons and shows a part of the process and the results. Although many of these workshops and masterclasses are focused on traditional craft techniques, they will inform the participant about the behaviour of a material.
Courtesy: The Institute of Making.

One of the unique aspects of the Institute of Making is its multidisciplinary profile, which is reflected in staff, members and the portfolio of projects. The staff are professionals with a varied background in craft, arts, engineering and material science. Most of them are both theoretically well accomplished with university degrees, but at the same time makers. They seem to bridge the identified gap between material science and art (Miodownik, 2003), but also between theory and practice. The members of the institute are faculty and students, not just from engineering and design, but from all departments of UCL. Projects such as 'Self-healing cities', 'Designing out plastic waste' and 'Nature inspired 4D printing for biomedical applications' and their many workshops are examples of how the Institute of Making brings together equipment, expertise and perspectives of making and materials from a wide range of disciplines such as craft, design, technology, history, philosophy, art, engineering and material science (Institute of Making, 2019).

One of the directors at the Institute of Making, Mark Miodownik, points out that they address problem solving by doing, not because theory is irrelevant to tackling complex problems, but because making stuff stimulates a creative approach (Miodownik, 2013). The institute's website says: 'We believe that until you make something you don't really understand it' – this is the basic, but valid argument for why their activities support teaching and research through making.

The four places introduced here are by no means the only ones suitable for material experimentation and making in educational institutions. Some long-standing design institutions, which rest on solid traditions of artistic practice and craft, like the Royal College of Art in London, have maintained a number of interesting workshops. However, these examples do show how different spaces for working with materials and making can be and how they are likely to reflect activities and educational programmes in their set-up, and perhaps in some cases also vice versa. Suitable physical facilities are essential for working with materials and making; but it is important to remember that it is not just a matter of installing a well-equipped lab or workshop. Tools on their own are not creative; thus for knowledge to be generated within the space in an educational setting, the support of leading faculty with contributory expertise and a solid foundation in experimental practice are required.

5.2 Learning to design with materials

The theory of learning is a broad and complex research field and it would take an entire separate book to cover all notions relevant for design education in depth. However, there are a few aspects that should be addressed, because they are fundamental for learning to design with materials. The didactic considerations introduced here relate to theories of cognitive processes presented in Chapters 2 and 3, which also addressed the significant connection between body and mind in the process of physically handling materials and tools. The knowledge required to design with materials involves not just formal knowledge and 'knowing about' but also a type of embodied knowledge of materials and manufacturing processes, which will enable the designer to practise creatively.

5.2.1 Different types of knowledge

Knowledge is both explicit and tacit. Explicit knowledge can be written down, captured in drawings and expressed in numbers, whereas tacit knowledge is related to skills in bodily movement and connected to the senses, physical experiences and intuition (Von Krogh, Ichijo & Nonaka, 2000; Atkinson & Claxton, 2008). A classic way to divide knowledge is separated into formal theoretical knowledge and the ability to practise. In reality this is obviously too simplistic a categorisation. Practice and theory do not have to exclude one another, and, as Dewey proposed almost a century ago, true theoretical knowledge not only arises through passive observation of reality, but may also arise through active partaking (Brinkmann & Tanggaard, 2010; Dewey, 1929). However, when discussing educational practice, the differentiation is useful considering that theoretical knowledge is related to 'knowing that or about' whereas practical knowledge involves 'knowing how'. This distinction, first made by philosopher Gilbert Ryle (Ryle, 1949), was later taken up – among others – by Tim Ingold, who addresses the meaning of knowing from the inside (Ingold, 2013).

In cognitive science 'knowing how' is described by some scientists as 'contributory' expertise. But, interestingly, they also operate with a category in between contributory expertise and formal theoretical knowledge called 'interactional' expertise (Collins, 2004; Collins & Evans, 2002; Collins, Evans & Weinel, 2016; Schilhab, 2007). One can learn 'interactional' expertise by immersing oneself in the terminology of a practical domain rather than the practice itself. The neurobiologist Theresa Schilhab describes how our mirror neuron system does in fact make it possible for someone who has never touched a hammer to be able to master the language of carpentry simply by being immersed for long enough in the carpenter's conversations alone. This way of learning would even provide the ability to converse expertly about the practical skill or expertise in carpentry, but still without being able to practise it (Collins, 2004; Schilhab, 2007). Although Ingold does not use this terminology he provides an example of what can be seen as the difference been interactional expertise and contributory expertise. He describes how studying everything about a cello player may enable one to write an insightful piece about playing cello, but learning to play the cello can only be done through actual playing of the cello (Ingold, 2013).

This more detailed classification of knowledge and expertise still suffers from the exclusiveness of separating and compartmentalising it. There may be many levels of contributory expertise and one critic points out that there is no clear cut between interactional and contributory expertise (Goddiksen, 2014). However, the distinction is useful for discussing the different types of knowledge necessary to design a physical sustainable product. In some areas the design graduate will need theoretical knowledge; in others interactional expertise may be valuable and, most importantly, it is necessary to consider the contributory expertise required to design. For example, whereas it may be 'enough' to have theoretical knowledge of the systems of sustainability, some level of interactional expertise in industrial manufacturing techniques could be useful, and, at least for a design process that includes a material dialogue, the designer would need to build contributory expertise in areas such as material experimentation and prototyping. No type of knowledge is by definition better than another, but the reason for differentiating them is that they require different didactic approaches.

5.2.2 Building contributory expertise

Learning theories have radically changed since the 1990s and although one can indeed learn from lectures and reading books, it would seem that the classic classroom setting with 'transmissive' instruction, where the information is believed to be transmitted from teachers to learners, may not even be ideal for transferring explicit knowledge (Land & Jonassen, 2014). The assumption with transmissive instruction is that if teachers communicate effectively to students what they know, the students will know it as well. However, based upon a compilation of contributions from experts within the field, Susan Land and David Jonassen point out that although transmissive instruction remains a popular way of teaching, there appears to be a consensus among researchers that learning is not a process of passive knowledge acquisition, but primarily a process of 'meaning making' that involves a wilful, intentional, active, conscious, constructive practice, which includes reciprocal intention-action-reflection activities (Land & Jonassen, 2014).

This is important to reflect upon as it questions the best way for students to acquire knowledge in general. Research on embodied cognition within the learning environment suggests the value of using symbolic materials, probes or models – even for acquiring explicit formal knowledge (Black, Segal, Vitale & Fadjo, 2012; Henriksen & Deep-Play Research Group, 2018). This is relevant for design education, but the focus here is on the *contributory knowledge* required to design with materials and how this can be acquired by a student in an educational setting. The reason being that this kind of knowledge *cannot* be learned through lectures and books alone, simply because it requires full-scale practical immersion within a specialist area of activity in order to gain the bodily knowledge of contributory expertise (Collins, 2004; Schilhab, 2007).

Theories of embodied cognition draw on results from neuroscience to explain underlying cognitive processes and thereby provide what may to some be a more precise explanation of how our hands, senses and the motor system are involved in the process of working with materials and tools. However, this way of understanding the connection between body and mind in a learning process is not entirely new; 80 years ago John Dewey wrote about 'learning by doing' and reasoned that integrating learning with experience made it more meaningful and usable (Dewey, 1938), and even earlier the educational philosophy of María Montessori, which also inspired Johannes Itten's foundation course at the Bauhaus, suggested that physical movement and touch would enhance learning (Lillard, 2016). Likewise it is reflected in Tim Ingold's justification of the value of 'knowing for one self' (Ingold, 2013) and in the theories presented by the architect Juhani Pallasmaa in the book *The eyes of the skin* (Pallasmaa, 2012), where he describes how it takes all of the senses to truly grasp materiality, or in his book *The thinking hand*, in which he argues that it is in the unity of mind and body that craftsmanship and artistic work can be fully realised (Pallasmaa, 2009).

Some academics may think of practical knowledge and skills in making as merely vocational or technical, but the study of craft rapidly shows a strong connection between body and mind. This also means that even though contributory knowledge is built in the individual through practice and experience, what is gained in the process is not just practical knowledge, but also tacit knowledge – and potentially even theoretical knowledge (of which the last decade of practice-based design research is a good example). Tacit knowledge was first defined by the polymath Michael Polanyi in 1966 (Polanyi, 1966) and is the embodied personal knowledge derived from experience and closely related to practice, skills and expertise. However, it also holds a dimension of more personal and subjective knowledge that is deeply rooted in an individual's actions and experiences and, thus, is very hard to formalise. This includes ideas, subjective insights, values and intuition (Atkinson & Claxton, 2008; Nonaka & Konno, 1998).

Several researchers have addressed tacit knowledge as central for design and craft expertise (Cross, 2006; Dormer, 1997; Reber, 1989; Rust, 2004). Unlike explicit knowledge, tacit knowledge resides only in people and, therefore, must be transferred from person to person. Thus, if knowledgeable people fail to pass on their tacit knowledge, it may disappear and be difficult to regain. Therefore, it also makes very good sense that traditionally crafts have been learned through apprenticeship, with master craftspeople transferring both their skills and tacit knowledge to the apprentice through demonstration and practice.

Engaging in a design process that includes a material dialogue requires contributory expertise, particularly in material experimentation and iterative prototyping. It is a type of know-how that includes both practical and tacit knowledge. In order to build this expertise within the individual student it is primarily a matter of what Pallasmaa describes embodied muscular mimesis acquired through practice and repetition, rather than conceptual or verbalised instruction (Pallasmaa, 2009). A more radical study, by Camilla Groth et al., showed that she was able to teach deaf-blind people how to throw clay by demonstrating with the touch of her hands alone (Groth, Mäkelä & Seitamaa-Hakkarainen, 2013). This study indicates that at least some practical and tacit knowledge might be transferred almost entirely without words.

Consequently, a substantial part of teaching a material dialogue can only be done through demonstration and practice – as it would be in an apprenticeship.

Nevertheless, it is essential to point out that this way of designing still requires a solid theoretical foundation and must entail reflective practice (Schön, 1983). Contributory expertise in material experimentation and iterative prototyping is essential for this way of designing, but it is of course only relevant if the designer also has the ability to contextualise the knowledge extracted from the material dialogue. This is evident, not least in relation to sustainability, where a designer must have the ability to operate at a product level, but at the same time understand the effect that the product will have on a systems level.

It may not be particularly difficult to appreciate the justification for transferring both tacit and practical knowledge through demonstration and practice, but depending on the individual design institutions, it may represent a challenge if this way of designing is to be implemented. I have already described the requirements for the physical spaces in the previous section, but depending on the profile of the staff, it could also affect the faculty. The reason is that whereas a lecturer with interactional expertise on a subject may contribute with theoretical knowledge, it will require contributory experts to educate designers with contributory expertise.

If the leading staff, such as the course director, the professors and the lecturers, primarily possess theoretical knowledge, are interactional experts or if their contributory expertise mainly lies in fields such as historical research, design strategy, anthropological studies, business planning or design thinking, and the only staff with contributory expertise in making are the technicians in a workshop, one could speculate that this may sustain and propagate a dualistic understanding of design activity (where design is understood as having two distinct and separate modes: one primary mode that contains the immaterial, the intellectual, abstract, creative, theoretical and conceptual and another, secondary mode that contains the practical, technical and concrete material reality). This could be further reinforced if the educational activities primarily take place in a space unsuited to working with materials and making.

To introduce a design process that includes a material dialogue will require staff with contributory expertise in design and making who are able to demonstrate and pass on the practice and the meaning of material experimentation and prototyping. Peter A. Hall from Central Saint Martins, University of the Arts London, contends that in order to solve complex problems, it is crucial for design education to achieve a balance between 'know-how' and meta-level thinking (Hall, 2016). This is absolutely true, but it is worth remembering that this balance can be present in an individual. A design educator is unlikely to be an all-in-one craftsperson, artist and materials scientist at the same time and the members of a faculty will and should have their individual competencies and professional specialisations. Nevertheless, a contributory expert can have a solid theoretical foundation, and considering the complexity of contributing to the field of sustainable design, it is definitely desirable. An example of this is the staff at the Institute of Making at UCL, who would all appear to have contributory expertise in making, but at the same time have a solid theoretical foundation from studies of material science, art or technology

The faculty of a design programme should combined represent the type of expertise and knowledge that the institution intends for the graduate to acquire. Yet, to properly reinstate working with materials in design education and to learn the meaning of a material dialogue would require a considerable part of the leading faculty to have contributory expertise related to designing with materials, making and experimental practice. Furthermore, the students would have to encounter this expertise within suitable physical facilities, such as laboratories or workshops where the student is supported in the process of acquiring contributory expertise through demonstration and practice.

5.2.3 Transferring the meaning of a material dialogue through experience

Including the material from the beginning can feel like a design process that has been turned upside down (Van Bezooyen, 2013). This represents quite accurately the initial reaction I have met from many design students, designers and faculty when they are first introduced to a design process that includes a material dialogue (Bak-Andersen, 2019). Still, if they have been taught and have practised a design process where the material reality of a design is a secondary consideration, separate from the conceptual, imaginative and immaterial ideation, their reaction is perhaps not surprising.

Although a review of the history of design education shows that materials have been central in the design process before and there are indications that interest in materials and making in design education is growing, a design process which includes a material dialogue may currently, in some places, break with present established understandings of the design process. A prevalent consensus on what is involved in design activity may be reflected in the curriculum, among the faculty and consequently settled as a 'default' mode for designing in the students. This represents an obstacle for teaching students to design with materials, because when they are challenged by a lack of skills or inadequate knowledge, they may return to this 'default' approach to the design process (Bak-Andersen, 2018). Consequently, the feeling that it is unnatural to design with materials can be a mental barrier for learning to design in this way and on a deeper level require a kind of 'reprogramming' of the student's understanding of the design process.

As already addressed the dilemma of this book is that it can only share theories, terminology and technicalities of designing with materials; it does not provide the reader with the practical or tacit knowledge to practise. Furthermore, although Chapter 4 outlines the elements of a material dialogue in some detail, it cannot be followed as a prescriptive recipe, simply because design thinking and making are never a simple linear process. As Buchanan stated, 'the problems addressed by designers do not, in actual practice, yield to any linear analysis and synthesis yet proposed' (Buchanan, 1995, p. 14). Additionally, the act of 'being in a dialogue with a tool and material' may sound entirely abstract to a student (or any other person) with no or limited experience of working with tools and materials, and they are unlikely to grasp the meaning of a material dialogue from reading or through verbal instructions. Consequently, the answer is, at least initially, not to read this book, but to learn the meaning of a material dialogue through experience.

Acquiring a fundamental understanding of what it means to be informed, restricted and inspired by a material and a tool can be learned through experience. A basic level of this embodied knowledge can be obtained through simple exercises where the material, the tool and even the product which the student is asked to produce are predefined. In the example presented here, the students were given a knife, an axe and a piece of wood in the morning and asked to carve a spoon by the end of the day. The students were enrolled in the Master's in 'Design for Emergent Futures' at the Institute of Advanced Architecture of Catalonia' in Barcelona. The exercise took place at the Green Fab Lab in Valldaura in February 2019 (Bak-Andersen, 2019).

Obviously, some tools and materials are easier to handle and manipulate than others, but for the student to have sufficient time to go into depth with the exercise and achieve an embodied understanding of the process will require some time. This particular exercise lasted nine hours, but more time could easily have been used. The instructions they received before starting were merely of a technical and practical nature. They were guided in how to use the provided tools through practical demonstration. Some instructions were given in plenum and some individually, typically involving showing them how to handle the tools in the most efficient and safe way. Apart from standing by with practical advice and a first aid kid, my colleague Thomas Duggan and I left the students on their own, and simply sat down next to them and carved. The students worked in an intense and very concentrated way for many hours, and the task of carving a spoon automatically entered them into a dialogue with the material (Figure 5.16).

In an exercise like this, the educator can observe the learning process of developing basic carving skills and see how the students will gradually gain control over the tool. Naturally, what is more difficult to detect through observation in such an exercise is the student's potentially growing embodied understanding of a material dialogue and the creative inspiration derived from the process. However, this tacit knowledge emerged at the end of the carving exercise, when all the students gathered with their spoons and were each asked to describe their process and share their experience.

As can be seen in Figure 5.17, the artefacts' worth as functional spoons was rather questionable, but as epistemic artefacts that provide knowledge and the bases for discussing the experience and meaning of a material dialogue, they were very valuable. During this session for reflection, it was evident that the knowledge acquired by the participants during the exercise was not only of a technical character. In this session they described how their interaction with the material and tool in combination with their deliberate intentions, actions and reflections during the process helped them to construct new knowledge and spark ideas.

When reflecting on the process, some would explain how they were directly inspired or restricted by the material when encountering a knot, a split or even when finding termites. The students mostly described the beginning of the

Figure 5.16 **A carving exercise, the process**

Students from the Master in 'Design for Emergent Futures' spend a day carving spoons and learning the meaning of a material dialogue at IAAC's Green Fab Lab in Valldaura, Barcelona.

Photograph: Thomas Duggan (Top).

Figure 5.17 **A carving exercise, the spoons**

All the spoons were presented at the end of the day and worked as epistemic artefacts and telltales in a discussion about design processes and the designer's control – or lack of control – over a design and its production.

Photograph: Thomas Duggan.

process as frustrating and difficult; some were afraid of injuring themselves with the tools and they felt that the tool and the material were working against them. However, nearly all of them also explained how they experienced a rapid improvement in their carving skills and before long they reached a stage where they felt like they were 'switching off' and relaxing. They would be carving, but at the same time thinking of other things, letting the mind wander. Several described this as a feeling of flow. Considering that all but one had no prior experience of carving wood it is interesting that the students would appear to have relatively quickly reached a level of automaticity that allowed them to handle the carving 'offline' (as defined by Wilson 2002, and described in Chapter 3).

The craftsmanship of the spoons was in general somewhat dubious, but still the students appeared very attached to them. Several said that they were proud of their spoon and would explain why it or the piece of wood was special, referring to aspects such as the depth of the colour, a pattern in the grain or the smell of a spoon made from laurel wood. They had clearly built a detailed sensorial understanding and appreciation of the material (which is very typical and can happen even to students working with culturally unacceptable or sensorially disagreeable materials such as human hair or sludge).

Finally, an interesting part of the conversation was related to control. The students were asked to reflect on why their spoons looked the way they did: which features they had explicitly decided on as designers and which had been defined by the materials and the tool, or could be seen as a reflection of their skills in carving. Considering the craftsmanship and the finish of the final spoons it is perhaps not surprising that the students felt that the material and their ability to use the tool had a considerable influence on the final result. Nevertheless, an interesting result was a conversation among the students about how much they would control in what they understood as a 'normal' design process. Several of the students had notable experience with digital design and manufacturing and whereas they previously had believed that they were entirely in control of the final prototype, they now started discussing if this was actually the case. The material would be restricted and defined by the tool, such as a 3D printer, and the design was not just the product of their free imagination, but also to some degree an expression of the software used and the manufacturing process.

Of course, this kind of exercise is not enough to learn to design with materials as described in Chapter 4. But it does provide something very central to this way of designing, namely an insight into the meaning of exploring a material and developing a design based on the results. It is a way for the student to acquire a fundamental embodied understanding of the meaning of being in a dialogue with a material and a tool, which is otherwise something that can be difficult for the student to grasp.

5.3 From analogue to digital and back

Considering the current level of digital design and advanced manufacturing technology, it is hard to imagine a design process, or any modern industrial fabrication process, that does not include at least some level of digitalisation. Therefore, when introducing the material dialogue from craft into a contemporary design process it will almost inevitably involve digital design and fabrication. The widespread use of digital tools makes treating the analogue and the digital as two separate modes of designing seem somewhat contrived. However, as the type of information retrieved from digital tools is different from that which can be retrieved from the use of analogue tools, it does, in this context, make sense to understand their separate values, as well as how their use is interlinked.

The mind and the body are inextricably linked and details of how this connection affects the way we learn and acquire knowledge has already been outlined. As described in Chapter 2, on the role of materials in design education, the design process – as taught in design schools – appears to have experienced a general shift from the hands to the head and from analogue to digital. There can be little doubt that digital design and fabrication are here to stay, but digital tools are different from most manual tools because in principle they can be controlled without physically experiencing the resistance of material. This means that digital tools will not provide the same level of embodied cognition of materials as manual tools, which tend to require a closer connection to the material being manipulated. Nevertheless, digital design and fabrication have other advantages and therefore it is relevant to address the constituents of a material dialogue in the context of digital design and fabrication.

5.3.1 Material knowledge in digital design

The term 'digital design and fabrication' typically refers to computer-aided design that is destined for different types of additive manufacturing, more popularly called 3D printing, or cutting machines such as CNC milling or laser cutting. Of course, this does not mean that all other design and fabrication processes are done manually: digital design is also used in many other fabrication processes, such as mould making or 3D knitting. Thus, whenever the designer is moving beyond conceptual design and developing a design for production, some level of digital design and fabrication are likely to be involved.

What makes it relevant to discuss digital design and fabrication within a material dialogue is that certain processes, such as additive manufacturing, are quite autonomous and require minimal human intervention during production (Muthu & Savalani, 2016). The design can be imagined and materialised without the designer physically handling or manipulating the material during the process. There will still be knowledge of materials involved, but unless the designer has prior embodied knowledge from working with a material, it will be limited to data that can be found on materials provided by the software. Perhaps for this reason, students, like several of the students who participated in the carving exercise, would describe themselves as fluent or very good at digital design and fabrication, but would at the same time declare that they had little or no knowledge of materials.

Exploring and manipulating a material will inform, restrict and inspire the designer. However, to extract this type of information demands physical interaction with the material by the designer in a way that involves virtually all senses. Without the touch of the hand, either directly or through tools that involve motor skills and feeling the resistance of the material, we cannot fully understand the material, and thus the designer who has never worked with materials may be inclined to concentrate on form. Digitalisation enables us to work with incredibly complex forms and advanced fabrication processes; nevertheless, on its own, it brings us no closer to an embodied understanding of the material.

Digital fabrication is advancing fast and can produce multifaceted forms in a wide variety of materials with a precision that would be hard to replicate by hand. However, it would appear that it is only recently that the more advanced labs and designers are starting to challenge the material limits of 3D printing. An example of an architect using 3D printing in a material-based process is Neri Oxman. In her doctoral thesis, 'Material-based design computation', she described her process as nature's way of designing and building, one in which material always precedes shape. In this she described how early forms of craft, as well as some of the most innovative new developments in materials science and engineering, utilise a material-based approach (Oxman, 2010). While drawing upon research on materials and biomimicry, her use of materials during her early projects seemed to be primarily defined by the limits of digital fabrication. However, Oxman and her team at the Mediated Matter Group at MIT have started to design materials and adapt printers in order to suit the material. An example of this is the 3D printer adapted for chitin paste made from a large quantity of crustacean shell waste (Mogas-Soldevila & Oxman, 2015).

Rivka Oxman, another architect, writes about the growing interest in materials-based design within digital design and fabrication and states that,

> processes of rapid prototyping employing fabrication technologies are an integral component in design and were being recognized as a significant technology that supports the full spectrum of digital design as a paperless process integrating computational models of generations and manufacturing in a single process'. The result of this is that, 'the architect and other design professions are regaining an important degree of knowledge and control of material and material processes.
>
> (Oxman, 2012, p. 428)

Without a profound study, it can be difficult to determine how particular experts have acquired their knowledge of materials, especially as tacit knowledge can be acquired independently of conscious attempts to do so (Reber, 1989) and many traits may be learned as a professional after graduating. Nevertheless, the type of knowledge required to design and develop a material for a 3D printer is not something that is provided by the 3D printer, or by the software. It involves knowledge of experimental practice and the ability to manipulate and design materials, which can only be acquired through experience.

This makes it interesting to follow the academic research of craftspeople or designers with good manual making skills that move into digital design and fabrication, such as the ceramicist Flemming Tvede Hansen or the textile designers Jane Taylor and Katherine Townsend. Hansen has extensive experience of working manually with clay and similar substances, and therefore inevitably brings this embodied knowledge of the material into his digital design and fabrication processes (Hansen, 2010). As early as 1997, another ceramicist, Neal French, confirmed the importance of CAD-CAM (computer-aided design and manufacturing) technology within the ceramic industry, and he argued that it would become more central as it developed further. Nevertheless, he also concluded – based upon his own experience as a designer working in the industry for twenty years – that users of CAD-CAM were better able to exploit its potential, and the potential of clay as a material, if they had the tacit knowledge of modelling by hand and by eye (French, 1997).

In the same period Gillian Crampton Smith, professor at the Royal College of Art in London, explained in an interview with design researcher Jeremy Myerson, how more 'playing the piano' in computing was needed. There should be programmes for virtuoso players that allowed a level of expertise where one can 'play' without thinking about it (Myerson, 1997). Programs for digital design and fabrication have radically changed since then, and the digital skills of many designers have developed to such a degree that at least some of them would appear to have acquired the level of expertise that Smith demanded.

Nevertheless, Taylor and Townsend point out that in knitwear design, despite machine manufacturers' efforts to make technology and software more user-friendly, the digital interface remains a significant barrier to many designers. They describe this barrier as a skill gap between knitwear designers and the technology, and while they address the necessity for the designer to be able to work with digital fabrication, in order to retain control at the point of production, they also highlight that the designer who is capable of using digital design and fabrication will lose touch with the materiality of the design process, if the work is being done through a computer interface alone (Taylor & Townsend, 2014). Therefore, an entirely digital design process could create a knowledge gap between the designer and the material.

Although some 3D drawing software does provide data on a limited range of materials and fabrication, it would seem that designers and researchers who have reached a proficiency which allows for experimentation and working creatively with materials in digital fabrication processes, such as 3D printing or 3D knitting, have not just taken ownership of a specific technology or tool and learned to use it for themselves. Rather, they are relying upon prior experiential craft knowledge (Taylor & Townsend, 2014). Thus, the value of tacit and practical knowledge derived from manual techniques or basic machines would appear to still be necessary for working creatively with materials in digital design and fabrication. This balance between analogue and digital skills should ideally also be reflected within educational practice.

5.3.2 Educational context

Despite interest over the last decade in makerspaces with digital fabrication tools, and efforts to design and create them, both inside and outside of design schools, there is still limited documentation on the content and processes of learning within these facilities (Sheridan et al., 2014). Prices have decreased, but it can still be expensive to establish a workshop for digital fabrication; 3D knitting machines, advanced 3D printers, CNC milling machines and robotic arms are generally costly, and rely on technology that is rapidly surpassed by newer and more advanced versions. Furthermore, many of these machines require trained technicians to operate them. However, the most advanced machines may not always produce the most advanced and creative designs, provide the best learning environments or result in the most skilled design graduates.

During the 1990s, many workshops within design schools were closed down. Not much documentation is available on this, but a UK study on the effects of technological change in industrial and product design courses within higher education attests to how these changes led to objections from the teaching staff. The staff argued that even though the machinery in the workshops no longer enabled the students to simulate the increasingly sophisticated manufacturing processes in the industry, workshop practice and model-making skills would still provide know-how. The workshops were seen as central for an exploratory design process, because they would provide the student with an insight into form and materials, and consequently assist in problem solving and decision-making (Myerson, 1991).

It would perhaps seem the most obvious conclusion that more advanced equipment in a workshop would result in more advanced projects and designs. However, considering what has already been outlined on sparking creativity in Chapter 4, central aspects were the importance of one's skills matching the challenge and receiving immediate feedback on one's actions (Csikszentmihalyi, 1996), as well as the value of tinkering (Pedersen, 2008). If the machinery is too advanced and requires specialist training by the manufacturer to operate, it is likely that only technicians employed at the workshop will know how to properly utilise it. This means that the student is removed from the tool, and as a consequence will disregard the knowledge that could be actively constructed through experience. Furthermore, the student will not be able to tinker with the tool, simply because it requires a certain level of control over the tool to gain the autonomy that can lead to unhindered experimentation.

Multi-material 3D printers are opening up for a wider range of materials, 3D printing of biomaterials is making advances in tissue engineering (Chia & Wu, 2015), and a robotic arm with an extruder for clay can be adapted to suit other materials of a similar consistency. Nevertheless, for supporting a material dialogue within an educational context, a simpler, home-built printer that the design student can learn how to operate and use for experimentation may be more useful in some cases. This type of printer will likely not provide the detail or finish that could be expected of more advanced machines; however, it will educate the student about the material and its possibilities for fabrication, qualify the student to conduct a dialogue with a technician or a manufacturer as a peer on equal terms, as well as prepare the student for working with more advanced tools.

A well-equipped workshop is the dream for any maker; but tools, both manual and digital, are not creative in themselves, and will not generate innovative designs. Like tacit knowledge, creativity resides in people. This may at least partly explain what engineer and educational scientist Paulo Blikstein calls the 'the keychain syndrome'. He describes how students will use well-equipped digital labs as fabrication facilities, producing trivial things on the laser cutter, rather than exploiting their potential for invention or experimental practice. He suggests that this may well be related to valuing the product over the process and the students will therefore settle for well-polished products produced with minimum effort, as opposed to messy, complex and potentially 'ugly' projects (Blikstein, 2014).

During the Industrial Revolution, craft was seen as in opposition to modernity, but good craftspeople did not necessarily get de-skilled. Some practitioners became specialised and were not against technology, but rather they pushed it forward (Adamson, 2013). Likewise, some designers and researchers are challenging the possibilities of the present digital fabrication technologies (Mogas-Soldevila & Oxman, 2015; Oxman, 2010; Oxman, 2015; Taylor & Townsend, 2014). Nevertheless, testing the material limits of digital fabrication tools requires sufficient control of a machine or tool to hack, adjust and tinker; but it also involves suitable facilities and knowledge of how to manipulate and explore materials.

5.3.3 Material dialogues with digital tools

There is no reason why digital design and fabrication should not be part of a design process that involves a material dialogue, and perhaps a very experienced designer with an extensive embodied knowledge of materials and fabrication could base ideas and the design on their own experience, and thus be able to stay within a digital realm throughout the process. However, a digital process is not a total solution, nor an end in itself, and a digital tool alone will not provide embodied knowledge about a material. Therefore, the design process should ideally include a blend of traditional and digital methods that inform one another.

This section has already introduced the value of craft skills and experiential knowledge of materials as a base for digital design and fabrication. Blikstein describes this as building on familiar practices by adding a layer of expressive technologies. He argues that a digital fabrication lab that merges computation, tinkering and engineering has the potential to augment, rather than replace, familiar and 'powerful' practices (Blikstein, 2014). Thus, when crafting is placed at the centre of creative design practice, it can be applied to support and further the potential of more advanced technology (Taylor & Townsend, 2014). For example, knowing how to knit by hand will enable an understanding of the principles of a 3D knitting machine. In an educational setting this means that simple mould-making exercises in plaster and silicone can assist the understanding and ability to discuss and design for manufacturing processes that require more advanced moulding techniques, even for students with almost no prior experience with making.

On the other hand, learning about digital fabrication methods can also inform manual experimentation. The machine-mediated experience of working by hand may prime the designer to 'think digitally' and make the transition to digital production (Piper & Townsend, 2015). An example of this can be seen in Figure 5.18, where a student, who has experience with 3D printing, has started developing her material to suit the extruder for the robotic arm. She prototyped the fabrication process by letting her own arm imitate the robotic arm and representing the extruder with a syringe.

Taylor and Townsend describe an iterative way of working in which hand and machine inform one another. Taylor uses embodied knowledge derived from hand-knitting in order to understand how industrial machines create a 3D garment, and, in turn, the knowledge that she gains through programming and working with technicians will inform methods used to create shaping by hand (Taylor & Townsend, 2014). Similarly, textile designer Anna Piper has a background in hand-weaving. She operates within a framework that situates the maker at the centre of creative practice, where hand production and digital production are carried out in tandem (Piper & Townsend, 2015). This way of using digital design and fabrication tools is very suitable for fostering a material dialogue.

Both digital and analogue tools are valuable in a material dialogue. But as these tools provide very different information, it is important to evaluate which is the most suitable for each enquiry. Furthermore, it can be necessary and valuable to experience and explore the analogue version of a digital fabrication process in order to understand its potential fully. This is particularly important within an educational setting, where it would appear that the ability to operate a basic digital fabrication tool is more relevant than the presence of very advanced machines that can only be operated by trained technicians. This is simply because the knowledge that is constructed in the process of tinkering with the machine leads to an autonomy that can build one's ability to push the limits of the tool and experiment more freely.

If the Maker Movement is indeed to influence design education, it could potentially provoke a noteworthy change, simply because it is based on *doing*. Making would likely require some practical skills and emerging into the field of practice – what Collins et al. described as requirements for becoming a 'contributory' expert with the ability to contribute to a field (Collins, 2004; Collins & Evans, 2002; Collins, Evans & Weinel, 2016). Nevertheless, the question remains as to *how* the Maker Movement will be integrated in design education. If the only change is adding new equipment to the workshop while the 'making' is not reflected in the educational practice, it is questionable to what degree it will actually lead to graduates with 'contributory' expertise.

Figure 5.18 **Developing a material to suit the tool**
This material made from orange peel through a process of material experimentation was developed by a student from the Master in 'Design for Emergent Futures' at IAAC. The aim is to adapt the material to suit a specific extruder on a robotic arm. An example of adapting a material to suit an additive manufacturing tool by expert design practitioners can be seen in Figure 4.14, which shows the work Dros and Klarenbeek are doing with Breda University to develop filaments for 3D printers made from algae.

5.3.4 The 'Ba' of material dialogue

Ikujiro Nonaka writes about the concept of 'Ba' as a foundation for knowledge creation. It was originally introduced by the Japanese philosopher Kitaro Nishida and further developed by Shimizu (Shimizu, 1995). The Japanese word 'Ba' roughly translates as 'place' in English. However, it should not necessarily be understood as a physical space, but rather as a foundation which unifies the virtual, physical and mental spaces as a platform for knowledge creation. It is a place which holds the framework, the theoretical foundation and a concentration of resources and facilitates what Nonaka describes as the spiralling process of the creation of new knowledge generated in the interactions between explicit and tacit knowledge (Nonaka, 1994).

This chapter has been an attempt to describe central elements in the 'Ba' for teaching a design process that includes a material dialogue. As there are considerable variations between different design educations, it is not possible to provide a clear-cut and exhaustive list of the potential implications if working with materials were to be reinstated as central to the design process. However, we can conclude that:

- Basic making skills, as well as an understanding of manufacturing processes, are essential – also for understanding digital fabrication.
- Designing with materials demands suitable physical facilities set up to work with material experimentation and prototyping.
- Leading faculty must have contributory expertise related to designing with materials, making and experimental practice, and their teaching should primarily take place in labs and workshops.
- The practical and tacit knowledge involved with a material dialogue must be demonstrated and acquired by the student through practice and repetition.
- Understanding the constituents and potential of a material will require a basic knowledge of topics from the natural sciences, such as the science of materials, chemistry, biology or biotechnology. This should be reflected in the curriculum, the physical facilities and also in contributory expertise within these field represented in the faculty (or associated with the faculty).
- Digital design and production should be carried out in tandem with material experimentation by hand.

References

Adamson, G. (2013). *The invention of craft*. London, UK: Bloomsbury Academic.
Ashby, M. (2008). The CES EduPack database of natural and man-made materials. Retrieved from https://www.grantadesign.com/download/pdf/biomaterials.pdf.

Ashby, M. F., & Johnson, K. (2013). *Materials and design: the art and science of material selection in product design*. Oxford, UK: Butterworth-Heinemann.

Atkinson, T., & Claxton, G. (2008). *The intuitive practitioner: on the value of not always knowing what one is doing*. Berkshire, UK: Open University Press.

Bak-Andersen, M. (2013). Facing the material challenge. *DAMADEI. Design and Advanced Materials as a Driver of European Innovation Report*. Retrieved from http://www.damadei.eu/report/.

Bak-Andersen, M. (2018). When matter leads to form: material driven design for sustainability. *Temes De Disseny: Nueva Etapa*, *34*, 10–33.

Bak-Andersen, M. (2019). From matter to form: reintroducing the material dialogue from craft into a contemporary design process. PhD thesis, the Royal Danish Academy of Fine Arts, Schools of Architecture, Design and Conservation.

Bang, A. L., & Eriksen, M. A. (2014). Experiments all the way in programmatic design research. *Artifact: Journal of Design Practice*, *3*(2), 4.1–4.14.

Baumeister, D., Tocke, R., Dwyer, J., Ritter, S., & Benyus, J. M. (2014). *Biomimicry resource handbook: a seed bank of best practices*. Missoula, MT: Biomimicry 3.8.

Black, J. B., Segal, A., Vitale, J., & Fadjo, C. L. (2012). Embodied cognition and learning environment design. In D. Jonassen, & S. Land (Eds.), *Theoretical foundations of learning environments* (pp. 198–223). New York, NY: Routledge.

Blikstein, P. (2014). Digital fabrication and 'making' in education: the democratization of invention. In J. Walter-Herrmann, & C. Büching (Eds.), *FabLab: of machines, makers and inventors*. Bielefeld, Germany: Transcript Publishers.

Brandt, E., & Binder, T. (2007). Experimental design research. 10th conference. Paper presented at the International Association of Societies of Design Research: Genealogy, intervention, argument. Hong Kong, China, 12–15 November.

Brinkmann, S., & Tanggaard, L. (2010). Toward an epistemology of the hand. *Studies in Philosophy and Education*, *29*(3), 243–257.

Buchanan, R. (1995). Wicked problems in design thinking. In V. Margolin, & R. Buchanan (Eds.), *The idea of design*. Cambridge, MA: MIT Press.

Chia, H. N., & Wu, B. M. (2015). Recent advances in 3D printing of biomaterials. *Journal of Biological Engineering*, *9*(1), 4.

Church, G. (2018). *Bio academy*. Retrieved 03/01/2018, from http://bio.academ any.org.

Collins, H. (2004). Interactional expertise as a third kind of knowledge. *Phenomenology and the Cognitive Sciences*, *3*(2), 125–143.

Collins, H. M., & Evans, R. (2002). The third wave of science studies: studies of expertise and experience. *Social Studies of Science*, *32*(2), 235–296.

Collins, H., Evans, R., & Weinel, M. (2016). Expertise revisited, part II: contributory expertise. *Studies in History and Philosophy of Science Part A*, *56*, 103–110.

Cross, N. (2006). *Designerly ways of knowing*. New York, NY: Springer.

Csikszentmihalyi, M. (1996). *Creativity, flow and the psychology of discovery and invention*. New York, NY: Harper Perennial.

Dewey, J. (2007/1938). *Experience and education* (2nd revised ed.). New York, NY: Simon & Schuster.

Dewey, J. (1929). *The quest for certainty.* Oxford, UK: Minton, Balch.

Dormer, P. (1997). *The culture of craft.* Manchester, UK: Manchester University Press.

Fab Foundation. (2018). *FabAcademy.* Retrieved 03/01/2018, from http://www.fabfoundation.org.

Frayling, C. (2012). *On craftsmanship: towards a new Bauhaus.* London, UK: Oberon Books.

French, N. (1997). CADCAM and the British ceramics tableware industry. In P. Dormer (Ed.), *The culture of craft* (pp. 158–167). Manchester, UK: Manchester University Press.

Goddiksen, M. (2014). Clarifying interactional and contributory expertise. *Studies in History and Philosophy of Science Part A, 47,* 111–117.

Groth, C., Mäkelä, M., & Seitamaa-Hakkarainen, P. (2013). Making sense: what can we learn from experts of tactile knowledge? *Form Akademisk Forskningstidsskrift for Design Og Designdidaktik, 6*(2).

Hall, P. A. (2016). Re-integrating design education: lessons from history. Paper presented at the Design Research Society 2016, 'Future-Focused Thinking', Brighton, UK.

Hansen, F. T. (2010). Material-driven 3D digital form giving, experimental use and integration of digital media in the field of ceramics. Doctoral dissertation, the Danish Design School.

Henriksen, D., & Deep-Play Research Group. (2018). *The 7 transdisciplinary cognitive skills for creative education.* Cham, Switzerland: Springer.

Ingold, T. (2013). *Making: anthropology, archaeology, art and architecture.* Abingdon-on-Thames, UK: Taylor & Francis.

Institute of Making (2019). Sixth Year Report, 2018–2019, London, UK: Institute of Making, UCL.

Kääriäinen, P., Tervinen, L., Vuorinen, T., & Riutta, N. (2020). *The CHEMARTS cookbook.* Helsinki, Finland: Aalto University.

Koskinen, I., Binder, F. T., & Redström, J. (2008). Lab, field, gallery, and beyond. *Artifact: Journal of Design Practice, 2*(1), 46–57.

Krogh, P. G., Markussen, T., & Bang, A. L. (2015). Ways of drifting: five methods of experimentation in research through design. In A. Chakrabarti (Ed.), *ICoRD'15: research into design across boundaries, volume 1* (pp. 39–50). New York, NY: Springer.

Land, S., & Jonassen, D. (2014). *Theoretical foundations of learning environments* (2nd ed.). New York, NY: Routledge.

Larsen, S. N. (2010). Den plastiske hjerne som mulighedsorgan – en analyse af samtidens smag for 'neuroplasticitet' og dens grænser. *Dansk Sociologi, 21*(2), 81–102.

Lillard, A. S. (2016). *Montessori: the science behind the genius.* Oxford, UK: Oxford University Press.

Miodownik, M. (2003). The case for teaching the arts. *Materials Today, 6*(12), 36–42.

Miodownik, M. (2013). The institute of making. *Materials Today, 16*(12), 458–459.

Mogas-Soldevila, L., & Oxman, N. (2015). Water-based engineering & fabrication: large-scale additive manufacturing of biomaterials. *MRS Proceedings 1800.*

Muthu, S. S., & Savalani, M. M. (2016). *Handbook of sustainability in additive manufacturing*. New York, NY: Springer.

Myerson, J. (1991). *Technological change and industrial design education*. London: Council for National Academic Awards.

Myerson, J. (1997). Tornadoes, T-squares and technology: can computing be a craft? In P. Dormer (Ed.), *The culture of craft* (pp. 176–185). Manchester, UK: University of Manchester Press.

Nonaka, I. (1994). A dynamic theory of organizational knowledge creation. *Organization Science*, *5*(1), 14–37.

Nonaka, I., & Konno, N. (1998). The concept of 'Ba': building a foundation for knowledge creation. *California Management Review*, *40*(3), 40–54.

Oxman, N. (2010). Material-based design computation. Doctoral dissertation, Massachusetts Institute of Technology.

Oxman, R. (2012). Informed tectonics in material-based design. *Design Studies*, *33*(5), 427–455.

Oxman, R. (2015). MFD: material-fabrication-design: a classification of models from prototyping to design. IASS 2015 Amsterdam Symposium: Future Visions – Performance Aided Design. pp. 1–11, Amsterdam, Holland: International Association for Shell and Spatial Structures.

Pallasmaa, J. (2009). *The thinking hand: existential and embodied wisdom in architecture*. Chichester, UK: Wiley.

Pallasmaa, J. (2012). *The eyes of the skin: architecture and the senses*. West Sussex, UK: John Wiley & Sons.

Pedersen, L. T. (2008). *Kreativitet skal læres! når talent bliver til innovation*. Aalborg, Denmark: Aalborg Universitetsforlag.

Piper, A., & Townsend, K. (2015). Crafting the composite garment: the role of hand weaving in digital creation. *Journal of Textile Design Research and Practice*, *3*(1–2), 3–26.

Polanyi, M. (1966). *The tacit dimension*. Chicago, IL: University of Chicago Press.

Pye, D. (1968). *The nature and art of workmanship*. Cambridge, UK: Cambridge University Press.

Reber, A. S. (1989). Implicit learning and tacit knowledge. *Journal of Experimental Psychology: General*, *118*(3), 219.

Rheinberger, H. (1997). *Toward a history of epistemic things: synthesizing proteins in the test tube*. Stanford, CA: Stanford University Press.

Rheinberger, H. (2013). Forming and being informed: Hans-Jörg Rheinberger in conversation with Michael Schwab. In Michael Schwab (ed.), *Experimental systems: future knowledge in artistic research* (pp. 198–219). Leuven, Belgium: Leuven University Press.

Rust, C. (2004). Design enquiry: tacit knowledge and invention in science. *Design Issues*, *20*(4), 76–85.

Ryle, G. (2009/1949). *The concept of mind*. London, UK: Routledge.

Schilhab, T. (2007). Interactional expertise through the looking glass: a peek at mirror neurons. *Studies in History and Philosophy of Science Part A*, *38*(4), 741–747.

Schön, D. A. (1983). *The reflective practitioner: how professionals think in action*. Abingdon-on-Thames, UK: Routledge.

Sheridan, K., Halverson, E. R., Litts, B., Brahms, L., Jacobs-Priebe, L., & Owens, T. (2014). Learning in the making: a comparative case study of three makerspaces. *Harvard Educational Review*, *84*(4), 505–531.

Shimizu, H. (1995). Ba-principle: new logic for the real-time emergence of information. *Holonics*, *5*(1), 67–79.

Steffen, D. (2013). Characteristics and interferences of experiments in science, the arts and in design research. Paper presented at the Nordic Design Research Conference, Malmö-Copenhagen.

Sutherland, T. D., Young, J. H., Weisman, S., Hayashi, C. Y., & Merritt, D. J. (2010). Insect silk: one name, many materials. *Annual Review of Entomology*, *55*, 171–188.

Taylor, J., & Townsend, K. (2014). Reprogramming the hand: bridging the craft skills gap in 3D/digital fashion knitwear design. *Craft Research*, *5*(2), 155–174.

Van Bezooyen, A. (2013). Materials driven design. In E. Karana, O. Pedgley & V. Rognoli (Eds.), *Materials experience:, fundamentals of materials and design* (pp. 277–286). Oxford, UK: Butterworth-Heinemann.

Von Krogh, G., Ichijo, K., & Nonaka, I. (2000). *Enabling knowledge creation: how to unlock the mystery of tacit knowledge and release the power of innovation*. Oxford, UK: Oxford University Press on Demand.

Wilkes, S., Wongsriruksa, S., Howes, P., Gamester, R., Witchel, H., Conreen, M., et al. (2016). Design tools for interdisciplinary translation of material experiences. *Materials & Design*, *90*, 1228–1237.

Wilson, M. (2002). Six views of embodied cognition. *Psychonomic Bulletin & Review*, *9*(4), 625–636.

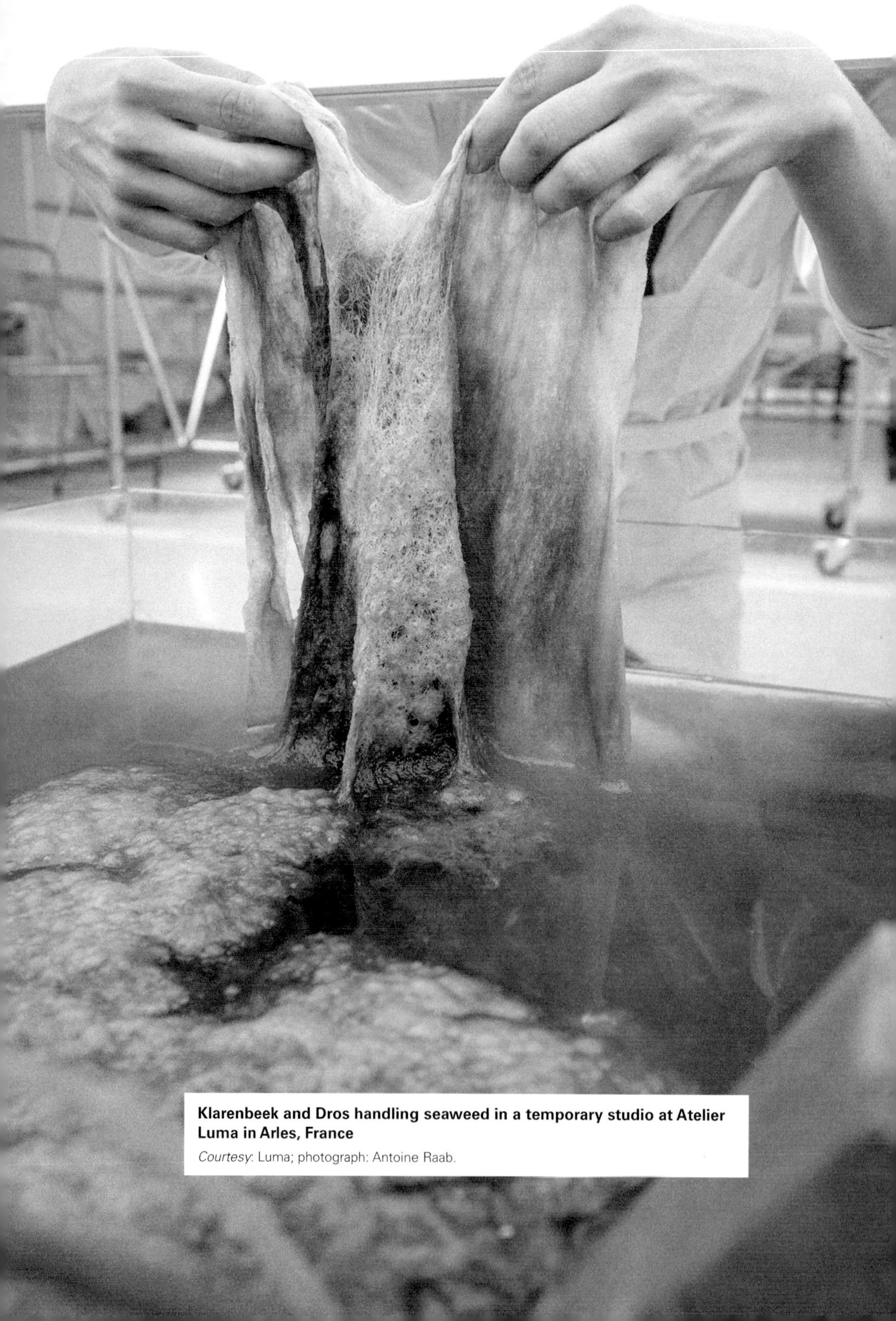

Klarenbeek and Dros handling seaweed in a temporary studio at Atelier Luma in Arles, France

Courtesy: Luma; photograph: Antoine Raab.

6 Sustainable design

Knowing how

There is little point in arguing with the obligation of designers to understand the system that they are designing for: the environmental crisis cannot be solved in isolation. This is essentially why a systemic approach to design for sustainability is considered better than alternative approaches, which merely focus on the product (Ceschin & Gaziulusoy, 2016). However, as has been described, there is a difference between 'knowing about' sustainable design and 'knowing how' to design for sustainability. Both ways of knowing can contain expertise relevant for sustainable design. Nevertheless, whereas *knowing about* is primarily based upon theoretical knowledge or knowledge *of* practice, and leads to interactional expertise or theoretical contributions, *knowing how* is based upon practical (and theoretical) knowledge and can lead to contributory expertise, which enables the designer to practise. Knowing how to design products according to the criteria for sustainability requires contributory expertise. This is why the main focus of this book has not been on systems of sustainability, but rather on the knowledge we need to design for sustainability and the educational practice involved.

Reintroducing materials into the design process and as central for the activity of design, can provide the designer with the material knowledge required to design for material circularity and other criteria such as minimising material consumption, energy consumption and toxic emissions, applying renewable and biocompatible resources into new designs, optimising the material in order to improve the lifespan of the materials and the products or knowing how to design for reuse, disassembly and recycling (Vezzoli & Manzini, 2008). These are all technical criteria, but equally important for sustainability is working with the meaning of materials and the experiential qualities that the materials will convey in a product (Karana, Hekkert & Kandachar, 2010; Karana, Pedgley & Rognoli, 2013). It is important to emphasise that knowledge of materials expressed in data or historical details can be very useful to the designer, but whereas memorising melting points, tensile strengths, statistics on availability and details about historical use does inform the designer about a material, it is a type of formal knowledge that will not in itself enable the designer to practise. To creatively design with materials, the designer must know how they can be manipulated and processed. This requires an embodied understanding of the behaviour and potential of a material, which can only be acquired and refined through experience.

Evidently, having the knowledge to design for sustainability does not by default ensure that the designer will indeed design a sustainable product. As will be introduced later in this chapter, it also requires a sense of responsibility and the intention to make the right decisions. Likewise, the book has addressed the material knowledge required to design for sustainability and how this is acquired, but it has not addressed the degree of sustainability that may be obtained by changing the way we design things or suggested the inclusion of any kind of tools for the measurement of sustainability. Whereas the environmental impact of a design in principle is quantifiable, knowledge about sustainability and materials does not necessarily lend itself to being expressed in numbers. Even so, to the readers who move in a professional environment primarily dominated by quantitative methods, to leave it out may seem deficient. Thus, it is worth sharing some considerations about the measurement of sustainable design and how this relates to innovation.

6.1 The dilemma of quantifying sustainable design

Considering that practically all fabrication of things will have a degree of negative environmental impact, tools for measuring the effect of a product, such as Life Cycle Assessment (LCA), are important for comparative analysis and for determining whether or not a design is indeed sustainable. LCA is by now a well-established tool to support science-based decision-making (Owsianiak, Bjørn, Laurent, Molin & Ryberg, 2018) and performing a full LCA of a product can be very relevant for designers. An LCA involves assessments of the environmental impacts associated with all stages of a product's life: from raw material extraction through to materials processing, manufacturing, distribution, use, repair and maintenance and, finally, disposal or recycling. All of these steps are in one way or another related to materials and the design of a product. Thus, it would perhaps appear to be an obvious tool for accompanying designing with materials.

However, although tools, such as the LCA, are developed to measure and advance sustainable design, in some cases, it may be ill-suited to advancing more profound aspects of sustainable innovation. On a fundamental level it is important to point out that not all aspects of sustainability can be measured. Had the shoes made from human hair in Figure 4.7 been put into production, they would most likely have achieved a brilliant score in an LCA, but the assessment would have said nothing of the acceptance of the user, the perceived value, the comfort or the aesthetics, which are equally important for the longevity and sustainability of a product. Furthermore, when engaging in a design process that includes a material dialogue, in principle all materials can be used, not just commercially available standard materials with elaborate data sheets. Through material studies and exploration the designer will gain a profound knowledge of the materials used, but if the material has recently come out of a laboratory, is a biomass waste material or is in another way a non-standard material and at present not utilised in industrial manufacturing it may prove difficult to get all the data required for a full LCA.

An LCA is expressed in numbers, but to achieve a precise result requires a precise input of data and this is not always available, particularly when a design is diverging from industry standards. To conduct a full LCA, extensive data is required, and as this can be difficult to obtain or may not even be available, some values must be based upon estimates (Gmelin & Seuring, 2014). This is one of the reasons why several LCAs on the same products might be based upon different figures and therefore come to a variety of conclusions (Van der Harst & Potting, 2013). Despite these uncertainties, LCAs are still expressed in exact numbers, but the result of an LCA is not necessarily unambiguous. The complexity in the results can be exemplified by looking into a comparative study using LCA to assess if a ceramic cup is more sustainable than a disposable paper cup (Martin, Bunsen & Ciroth, 2018). One may expect that the long-lasting ceramic cup is by far the more sustainable, but interestingly the study shows that in some cases the disposable paper cup is in fact more sustainable than the ceramic cup. When studying the details, it is clear that there is no single definitive answer. As in many comparative analyses of products, the answer is complex and depends upon many factors. In this case, the ceramic cup is indeed more sustainable, but only when it has been used at least 140 times, and only if it has been washed in a modern dishwasher. Very different results appear when the ceramic cup is washed by hand in warm water (Martin et al., 2018).

Particularly when trying to understand the value of biomass waste materials, it makes sense to look into material flow analysis, which is a systematic assessment of the flows and stocks of materials within a system defined in space and time (Brunner, Morf & Rechberger, 2002). This can be very useful in LCA and in general for understanding where the raw material comes from and where it will go at the end of a product's life. However, even in Europe where most countries publish economy-wide material flow accountings (EW-MFA), specific details of local waste-management systems are rarely available at a city and regional level (Laurent et al., 2014). This forces an LCA for a product made of a local waste material to be based upon very general estimates, and consequently the results become inaccurate. Furthermore, in a recent study on residual biomass as a resource, it has been pointed out that there are still some doubts concerning how to conduct LCA when using biomass waste as a raw material. The reason for this is that when waste is recirculated as raw materials within a resource system, it is in most cases automatically considered 'zero-burden' (Olofsson & Börjesson, 2018). This can produce results that are misleading in a comparative analysis with standard materials, but it also makes an LCA on products made from non-standard materials, such as biomass waste, into a difficult guessing game of numbers with results that at best can be considered indicative.

When precise data is available an LCA is a critical tool to evaluate the sustainability impact of different materials, production methods, transportation alternatives, etc. But it is a tool that should be used with respect for its limitations. As already stated, not all aspects of sustainability can be measured: aspects such as emotional attachment, meaning of materials and perceived obsolescence are difficult to quantify. But more specifically, the LCA as a tool within a design process that deals with the more innovative and fundamental aspects of sustainable design may not be an ideal match. It is not necessarily well suited for working with untried materials, such as biomass waste or materials that have recently come out of labs. It is also questionable whether it is an appropriate tool for evaluating new fabrication technologies, simply because these will rarely have been optimised in the same way as more established ones, and, thus, in an LCA they may appear less sustainable than those which they are intending to replace. However, some of these emerging materials and new fabrication technologies represent considerable potential for sustainability in the future (Muthu & Savalani, 2016).

An example of this is additive manufacturing, which is a rapidly developing fabrication technology. An exploratory study of LCAs of additive manufacturing technologies and additively manufactured products suggests that LCA can be used as a tool to improve a fabrication technology. However, the study also emphasises the importance of the decisions that are made at the design stage and proposes that to fully evaluate and reveal the sustainability advantages of additive manufacturing, a more holistic assessment framework than LCA is needed (Ma, Harstvedt, Dunaway, Bian & Jaradat, 2018). On the relationship between LCA and innovation, a recent paper concludes that in principle the convergence of 'eco-innovation' and LCA studies is plausible, but states that the literature uniting both themes is scarcely found in publications within the area of innovation. Among the differing suggestions as to the reasons for this, the authors focus on a possible conflict in the skills required. They argue that innovation requires virtues such as creativity, risk-taking, collaboration and knowledge gathering, whereas a typical professional conducting an LCA has to master methodologies and management tools, as well as the ability to perform analytical tasks (Motta, Issberner & Prado, 2018).

Consequently, an LCA can be a very useful tool for measuring sustainability, but it is not necessarily suitable for aspects related to sustainable innovation, as the LCA expresses only a quantifiable existing reality and not the future potential of something.

6.2 Co-creation and distributed knowledge

There may be a common belief that advanced technology can save us and solve the environmental crisis, but there is little to support that this is indeed true (Huesemann & Huesemann, 2011). Still, considering how the complexity and numbers of manufacturing processes have increased in the last 50 years and how the technology is constantly advancing, not least within digital design and fabrication, it is likely that the designer in the future will have to relate to a production chain defined by augmented complexity.

One of the interesting features of digital fabrication technologies such as 3D printing is the potential for direct production straight into end use. If the designer is able to operate the fabrication tools, he or she can in principle control the entire process. Considering some of the literature presented in the first chapters of this book on the presumed division of labour between imagining and making, this is interesting, because it promises to set right some of the adverse effects of the Industrial Revolution. Objections to the Industrial Revolution were not only a matter of protecting craft skills, as William Morris argued, but were also a question of protecting the measure of control the craftspeople exercised over their work (Frayling, 2012).

However, direct production, where the designer can be compared to the individual craftsperson who masters both design and manufacture, is only possible with relatively simple designs. No one person can make complex products, such as a car, or all parts for a washing machine, simply because the knowledge is spread across various professions (Dormer, 1997). A designer will almost inevitably, at some point, have to design products where the knowledge and skills required to complete the process are distributed between fields. Consequently, if the designer is to be more than the 'ideator' of a design and maintain a measure of control over its production, it will entail co-creation.

'Co-creation' is a broad term that refers to the act of collective creativity (Sanders & Stappers, 2008). It is often described more specifically as co-design or participatory design, which tends to be focused on the involvement of the future user in the design process (Binder et al., 2011; Simonsen & Robertson, 2012). The end user is equally important in a design process that involves a material dialogue – even during the early phases of material experimentation: for the meaning of materials is not a measurable quantity observable in the lab, but something that is culturally based; and it can determine acceptance or rejection of the material by a future user. Nevertheless, the co-creation that should be discussed here in the context of distributed knowledge in the production chain is not related to the user, but, rather, to the professionals who possess the knowledge and skills necessary to complete a design solution.

In 1956, Don Wallance wrote about the need for co-creation in design. Wallance was a recognised industrial designer in the USA. Like most of the designers behind Danish Modern, which occurred during the same period, Wallance had a background in craft. He was a metalworker and his understanding of the material was apparent in his designs. Nevertheless, he still highlighted how no individual in a large-scale industry is capable of creating a product alone and that a product must be created by activating and synchronising the thoughts and operations of many people (Wallance, 1956). Therefore, even a designer with a background in craft and a profound experiential knowledge of materials will need co-creators to design a complex product.

Although Pye in his book on craftsmanship (Pye, 1968) appears to imply that a good designer must understand all possibilities of a material for a design, he later admitted in an interview with Frayling that there are exceptions to this. He provides the examples of British designer Dick Russell and the Danish designer Kaare Klint, who produced excellent designs without a background in craft. By working very closely with the same skilled makers for the duration of their careers, they accomplished the same level of craftmanship and sensibility for materials in their results (Frayling, 2012). Designers need not have the same type of contributory expertise in making and handling materials as a craftsperson. However, in order to design for production and maintain control over the product, designers do need to be able to design with materials and this requires the ability to work in an experimental practice with materials, having basic making skills and an understanding of manufacturing processes. Every design problem requires a different solution, and therefore a different combination of advisors and co-creators. By including a material dialogue within the design process, the designer does not gain the skills to produce the design; however, they may acquire the knowledge to be able to ask informed questions of specialists such as material scientists, chemists, biologists, technicians, engineers and manufactures.

Digital fabrication technology is developing rapidly and although direct production does potentially give the designer control over the entire design and fabrication process, advancements in fabrication technology are in general more likely to distribute knowledge. Therefore, co-creation and collaboration with professionals from other fields is unavoidable for any designer who wishes to design for production. However, in order to avoid any 'skills gaps' which can obstruct the communication between designer and specialists, an understanding of the materials and fabrication processes in question is essential. Kääriäinen et al. from CHEMARTS at Aalto University describe how working side by side is the key to a successful collaboration between professionals from different disciplines. They point out that it should be approached in the same way as learning a new language and requires an open mind and heart, patience and plenty of practice (Kääriäinen, Tervinen, Vuorinen & Riutta, 2020).

6.3 The responsibility of designers

By introducing more human-made things into the system, designers carry a responsibility as re-creators of the world (Findeli, 2008), and even though designing with materials can inform designers about the physical realm of a product and can provide them with knowledge to design for sustainability, it is still up to the designers to make the right decisions – and this may not always be an easy task. Designers play a part in a dynamic whole, and the process of designing and fabricating a product often involves numerous stakeholders with diverse interests. In order to achieve sustainability, all stakeholders must operate within the same system and it is necessary to look at the interfaces and complex coupling between the entire system's components in order to understand the function and effect of the sum of its parts (Mauser et al., 2013).

Nevertheless, whereas this is the ideal and the future that we must strive for, in the present situation designers are still likely to be challenged in their capacity to make the right choices. Factors such as profit margin, the ease of maintaining standard manufacturing techniques or a resistance from technicians to using unfamiliar materials and binders may to some stakeholders be considerably more important than developing a sustainable product. This potential conflict of interest is not likely to entirely disappear, but considering how many companies are taking steps towards sustainable production or a circular economy, like IKEA's circular product design principles (IKEA, 2019), it does suggest the bases for a more common ground in the future.

It is equally important to point out that even good intentions and making what would appear to be the right choices is not always good enough. Even with all of the literature available on a subject and with expertise in designing, we can only practise within the limits of our current knowledge. Therefore, in spite of good intentions, we can still make mistakes. Frayling provides the example of Thomas

Midgley, a designer and a chemist who back in 1921 solved a knocking effect in car combustion engines by adding lead to the petrol with the best of intentions. It solved an immediate problem, but, as we now know, it had unforeseen consequences. Similarly, he replaced toxic chemicals in refrigerators with Chlorofluorocarbon (CFC) gases, which at the time appeared to be harmless, but later turned out to destroy the ozone layer (Frayling, 2012).

A more recent example that present product designers have been faced with is the use of bisphenol A (BPA), which is used in the production of high-volume polycarbonate and epoxy resin compounds and found in a number of consumer products that include plastic bottles and the linings of canned goods. Unfortunately, when BPA is used in food containers, a small amount migrates into the food and it has proved to be potentially damaging to the human body by causing disorders such as advancing puberty (Howdeshell, Hotchkiss, Thayer, Vandenbergh & Vom Saal, 1999). Therefore, many conscientious designers and manufacturing companies substituted it with bisphenol S (BPS) and sold their products as BPA-free. However, although BPS was once considered a safe substitute for BPA, it was later documented that bisphenol S is equally toxic (Ji, Hong, Kho & Choi, 2013; Vinas & Watson, 2013). Designers should not be discouraged from trying to do the right thing, but merely be aware that designing for sustainability adds complexity, which sometimes calls for close collaboration with material scientists, and in all cases calls for caution.

6.4 The future of sustainable design

At present we are designing products for an incomplete system of sustainability. Even in countries that are focused on solving the environmental crisis, the infrastructure, the legislation, the services, the facilities, the waste management and the range of certifications are still in the process of being defined and constructed. However, it is important to remember that when designing a product today, we are in fact designing a product for the future. Perhaps the product is produced and sold within a relatively short period of time, but it may be in use for many years and end its life in a much more distant future. Therefore, we should not just design for the minimum requirements of today's environmental legislation; the circularity of a design cannot be ignored, just because the present waste management system is not yet properly set up to handle it, and we must not wait for the users to demand that we design their products in a different way.

In 50 years' time, we could potentially be at a stage where materials are recycled or biodegraded in a system of closed loops, and we will have little or no material that is considered waste. Today, a large percentage of waste materials are mixed from both natural materials and an endless amount of different synthetic materials, which are hard or impossible to separate. It is difficult to make a material better than it was when it was first utilised in a product. At present, this means that some waste materials are so polluted or mixed that with today's technology they cannot be separated. These types of waste materials are not suitable as raw material for a product that has to be recirculated at the end of its life. Consequently, not all materials are suitable for sustainable design.

A design that is not informed by a material reality may be good conceptually or look attractive in a drawing, but a designer who does not know how to design with materials is in many respects as badly equipped as a chef who does not understand the ingredients for the dish she is preparing. Qualities such as innovation and sustainability are not additives that can be injected into a product at the last minute, meaning that a product not originally designed to fulfil these criteria is unlikely ever to do so. Designing for sustainability means taking responsibility for the designs that we contribute to the world. Design educators are obliged to instil a moral framework and build this sense of responsibility within their students. However, to take responsibility requires a certain level of control over the final design and production, and to exert that control requires knowledge about materials and manufacturing processes. Expertise in designing creatively with materials can enable designers to stay in control of the production process, or at least enable them to conduct a competent discussion with stakeholders and production personnel. In this way the designer is not just the creator of a concept but becomes the designer of a recipe to be followed in the production chain.

Another aspect that potentially changes the role of the designer is the approach to material selection. As Jonas Edvard states in the interview in Chapter 4, standard prefabricated materials, like rolls of woven textiles or board materials like MDF or plywood, are designed for standard production methods, but they do not necessarily offer the most efficient way to get from the raw material to a finished product; nor do they inspire innovative and new manufacturing processes. Including a material dialogue into a design process does not by default mean that the designer has to work with unwashed wool or raw timber. However, exploring less-processed materials may support the designer in finding a more efficient path from the raw material to a finished product. For instance, working with pieces of veneer instead of finished plywood or wooden fibres instead of MDF board will make it possible to press the material directly into the desired shape using a different type of binder. Likewise working with yarn or fibre, instead of fabric by the yard, will allow working with manufacturing technologies such as 3D knitting and thereby skipping the steps of making a roll of fabric that has to be cut up and sewn back together. By moving away from prefabricated materials and allowing any potentially suitable material to enter the design process opens up for a much larger and more varied set of resources and manufacturing techniques. In the process it alters the role of the designer slightly: in a design process where the material reality is secondary to concept and form, the designer is likely to be the selector of a standard 'off-the-shelf' material. In a design process that includes a material dialogue and less-processed materials, the product designer is not just a material selector, but to some extent also a material designer.

The design profession, as well as our understanding of the activity of design, are constantly evolving and will presumably continue to do so. External elements such as advances in technology, improved understanding of how nature builds, and the challenges posed by social, environmental and economic conditions will likely be reflected in the focus of the design process and force it to adapt. However, considering the present environmental crisis there can be little doubt that the way we currently design things has to change rapidly. Unfortunately, there is no book which contains all the answers for the future of sustainable design and production; like Dros and Klarenbeek experienced, we must educate ourselves. To learn from each other, designers – and architects, engineers and anyone else involved with designing and fabricating our built environment – must start sharing failures. On no account do I wish to diminish the accomplishments of professionals who are actually working with sustainable design, but to move forward we must also share the difficulties we encounter and not just our successes. The polished versions of our work, which look good in a design magazine, may inspire other designers, but by sharing the mistakes we make and the barriers we encounter in the process of designing, developing and manufacturing sustainable products, we expose flaws in the system, identify aspects in the way we design that it would be beneficial to alter and, most importantly, we have the chance to learn from each other and push forward sustainable design.

How we design and how we learn to design inevitably affects what we design. Therefore, I hope this book will not just inspire designers to learn about sustainability but also encourage them to start exploring the material reality of the physical world: to dive into materials and fall in love with them, to experiment and design with them. I hope it will motivate designers to collaborate with other disciplines and, not least, inspire those designers with spotless studios to turn their working environment into productive workshops or laboratories, where there is space to experiment and tinker with materials, tools and manufacturing processes, and where they potentially can build the knowledge that will enable them to design for sustainability. Equally, I hope it may contribute to a discussion about how we teach design and sustainability, and also to a more profound and complex discussion about the field of design, which could become more closely related to topics from the natural sciences, such as biotechnology, biology and chemistry, if materials are to be central in the activity of design again.

The motivation for writing this book is the determination to change the way we design things, but when reflecting upon how to design for sustainability, it is perhaps relevant to consider if the content would have been notably different if instead of sustainability the focus had been to design for issues such as cost reduction, improving quality, aesthetics or simplifying production processes. The knowledge of materials and how to work with them is not just a significant component for closing the gap between the criteria for sustainable product design and the knowledge and ability needed to design for these: it is fundamental knowledge for designers who wish to remain in control of their designs and the production of these. And thus, it may not merely describe the type of material knowledge required to design for sustainability, but more generally the material knowledge needed to creatively improve the design of physical objects.

References

Binder, T., Brandt, E., Halse, J., Foverskov, M., Olander, S., & Yndigegn, S. (2011). Living the (co-design) lab. Paper presented at the Nordes, 4th Nordic Design Research Conference.

Brunner, P. H., Morf, L. S., & Rechberger, H. (2002). *Thermal waste treatment: a necessary element for sustainable waste management.* Vienna University of Technology and Institute for Water Quality and Waste Management.

Ceschin, F., & Gaziulusoy, I. (2016). Evolution of design for sustainability: from product design to design for system innovations and transitions. *Design Studies*, *47*, 118–163.

Dormer, P. (1997). *The culture of craft.* Manchester, UK: Manchester University Press.

Findeli, A. (2008). Sustainable design: a critique of the current tripolar model. *The Design Journal*, *11*(3), 301–322.

Frayling, C. (2012). *On craftsmanship: towards a new Bauhaus.* London, UK: Oberon Books.

Gmelin, H., & Seuring, S. (2014). Achieving sustainable new product development by integrating product life-cycle management capabilities. *International Journal of Production Economics*, *154*, 166–177.

Howdeshell, K. L., Hotchkiss, A. K., Thayer, K. A., Vandenbergh, J. G., & Vom Saal, F. S. (1999). Environmental toxins: exposure to bisphenol A advances puberty. *Nature*, *401*(6755), 763.

Huesemann, M., & Huesemann, J. (2011). *Techno-fix: why technology won't save us or the environment.* Gabriola, BC, Canada: New Society Publishers.

IKEA. (2019). *Circular product design guide.* Retrieved 06/15/2020, from https://preview.thenewsmarket.com/Previews/IKEA/DocumentAssets/512088_v2.pdf.

Ji, K., Hong, S., Kho, Y., & Choi, K. (2013). Effects of bisphenol S exposure on endocrine functions and reproduction of zebrafish. *Environmental Science & Technology*, *47*(15), 8793–8800.

Kääriäinen, P., Tervinen, L., Vuorinen, T., & Riutta, N. (2020). *The CHEMARTS cookbook.* Helsinki, Finland: Aalto University.

Karana, E., Hekkert, P., & Kandachar, P. (2010). A tool for meaning driven materials selection. *Materials & Design*, *31*(6), 2932–2941.

Karana, E., Pedgley, O., & Rognoli, V. (2013). *Materials experience: fundamentals of materials and design.* Oxford: Butterworth-Heinemann.

Laurent, A., Bakas, I., Clavreul, J., Bernstad, A., Niero, M., Gentil, E., et al. (2014). Review of LCA studies of solid waste management systems – Part I: lessons learned and perspectives. *Waste Management*, *34*(3), 573–588.

Ma, J., Harstvedt, J. D., Dunaway, D., Bian, L., & Jaradat, R. (2018). An exploratory investigation of additively manufactured product life cycle sustainability assessment. *Journal of Cleaner Production*, *192*, 55–70.

Martin, S., Bunsen, J., & Ciroth, A. (2018). *Case study: ceramic cup vs. paper cup.* Berlin, Germany: OpenLCA.

Mauser, W., Klepper, G., Rice, M., Schmalzbauer, B. S., Hackmann, H., Leemans, R., et al. (2013). Transdisciplinary global change research: the co-creation of knowledge for sustainability. *Current Opinion in Environmental Sustainability, 5*(3–4), 420–431.

Motta, W. H., Issberner, L., & Prado, P. (2018). Life cycle assessment and eco-innovations: what kind of convergence is possible? *Journal of Cleaner Production, 187*, 1103–1114. Retrieved from http://www.sciencedirect.com.ez-kab.statsbiblioteket.dk:2048/science/article/pii/S0959652618308989.

Muthu, S. S., & Savalani, M. M. (2016). *Handbook of sustainability in additive manufacturing*. New York, NY: Springer.

Olofsson, J., & Börjesson, P. (2018). Residual biomass as resource: life-cycle environmental impact of wastes in circular resource systems. *Journal of Cleaner Production, 196*, 997–1006.

Owsianiak, M., Bjørn, A., Laurent, A., Molin, C., & Ryberg, M. W. (2018). LCA applications. In M. Z. Hauschild, R. K. Rosenblaum and S. I. Olsen (Eds.), *Life cycle assessment* (pp. 31–41). New York, NY: Springer.

Pye, D. (1968). *The nature and art of workmanship*. Cambridge, UK: Cambridge University Press.

Sanders, E. B., & Stappers, P. J. (2008). Co-creation and the new landscapes of design. *Co-Design, 4*(1), 5–18.

Simonsen, J., & Robertson, T. (2012). *Routledge international handbook of participatory design*. New York, NY: Routledge.

Van der Harst, E., & Potting, J. (2013). A critical comparison of ten disposable cup LCAs. *Environmental Impact Assessment Review, 43*, 86–96.

Vezzoli, C., & Manzini, E. (2008). *Design for environmental sustainability*. New York, NY: Springer.

Vinas, R., & Watson, C. S. (2013). Bisphenol S disrupts estradiol-induced nongenomic signaling in a rat pituitary cell line: Effects on cell functions. *Environmental Health Perspectives, 121*(3), 352–358.

Wallance, D. (1956). *Shaping America's products*. New York, NY: Reinhold.

Index